Multidimensional Systems Theory

Mathematics and Its Applications

Multidimensional Systems Theory

Progress, Directions and Open
Problems in Multidimensional Systems

Edited by

N. K. Bose

School of Engineering, University of Pittsburgh, U.S.A.

With contributions by
N. K. Bose, J. P. Guiver, E. W. Kamen,
H. M. Valenzuela, and B. Buchberger

Springer-Science+Business Media, B.V.

Library of Congress Cataloging in Publication Data

CIP

Main entry under title:

Multidimensional systems theory.
     (Mathematics and its applications)
     Includes bibliographies and index.
     1. System analysis. I. Bose, N. K. (Nirmal K.), 1940– . II. Guiver, J.
P. III. Series: Mathematics and its applications (D. Reidel Publishing Company)
QA402.M83   1984   003   84–15060
ISBN 978-1-4020-0328-8   ISBN 978-94-009-5225-6 (eBook)
DOI 10.1007/978-94-009-5225-6

Table of Contents

Editor's Preface

Approach your problem from the right end and begin with the answers. Then one day, perhaps you will find the final question.

The Hermit Clad in Crane Feathers in R. van Gulik's The Chinese Maze Murders.

It isn't that they can't see the solution. It is that they can't see the problem.

G. K. Chesterton. The Scandal of Father Brown The point of a Pin.

Growing specialization and diversification have brought a host of monographs and textbooks on increasingly specialized topics. However, the "tree" of knowledge of mathematics and related fields does not grow only by putting forth new branches. It also happens, quite often in fact, that branches which were thought to be completely disparate are suddenly seen to be related.

Further, the kind and level of sophistication of mathematics applied in various sciences has changed drastically in recent years: measure theory is used (non-trivially) in regional and theoretical economics; algebraic geometry interacts with physics; the Minkowsky lemma, coding theory and the structure of water meet one another in packing and covering theory; quantum fields, crystal defects and mathematical programming profit from homotopy theory; Lie algebras are relevant to filtering; and prediction and electrical engineering can use Stein spaces. And in addition to this there are such new emerging subdisciplines as "experimental mathematical", "CFD", "completely integrable systems", "chaos, synergetics and large-scale order", which are almost impossible to fit into the existing classification schemes. They draw upon widely different sections of mathematics. This programme, Mathematics and Its Applications, is devoted to new emerging (sub)disciplines and to such (new) interrelations as exempla gratia:

–a central concept which plays an important role in several different mathematical and/or scientific specialized areas;
–new applications of the results and ideas from one area of scientific endeavour into another;
–influences which the results, problems and concepts of one field of enquiry have and have had on the development of another.

ix

The Mathematics and Its Applications programme tries to make available a careful selection of books which fit the philosophy outlined above. With such books, which are stimulating rather than definitive, intriguing rather than encyclopaedic, we hope to contribute something towards better communication among the practitioners in diversified fields.

As mathematical disciplines go, system and control theory is a young one. Definitely post World War II though, of course, its roots go back further. It has already established itself as a determined user of the most sophisticated mathematical tools available such as algebraic geometric intersection theory and cohomology, all kinds of spectral factorizations, various chapters in interpolation theory, H^2 and H^∞ spaces, ... It also, concomitantly, seems to be an inexhaustible source of new and interesting problems.

Some of these problems have to do with different aspects (than usually considered) of well-established areas of concern.

Examples are: is there an effective (constructive) proof of the Quillen–Suslin theorem on algebraic vector bundles, and what happens to the idea of a conservation law if imputs are allowed?

The lines just written are true even for the usual kind of systems, or 1-D systems, and in fact even for linear systems which at first sight seem so simple (mathematically speaking) that it is hard to believe that sophisticated tools will be needed to deal with them.

Modern technological demands such as for image processing and the remarkable successes of 1-D theory (i.e. the usual kind of mathematical system theory) caused the emergence of 2-D (and n-D) theory. This is a field in which things are in rapid flux calling for an occasional taking stock of the situation. That is exactly what this book aims to do.

The unreasonable effectiveness of mathematics in science . . .

Eugene Wigner

Well, if you know of a better 'ole, go to it.

Bruce Bairnsfather

What is now proved was once only imagined.

William Blake.

As long as algebra and geometry proceeded along separate paths, their advance was slow and their applications limited.
But when these sciences joined company they drew from each other fresh vitality and thenceforward marched on at a rapid pace towards perfection.

Joseph Louis Lagrange.

Bussum, March 1985   MICHIEL HAZEWINKEL

Preface

Towards the end of 1981, Professor M. Hazewinkel invited me to write a book on multidimensional systems theory in his series devoted to mathematics and its applications. At that time, however, I had already completed the manuscript for the text entitled, 'Applied Multidimensional Systems Theory', which was published by Van Nostrand Reinhold Co. during the first quarter of 1982. Though, at that time, the need for a textbook on the subject was realized, it was felt that the area of multidimensional systems theory was expanding at a prolific rate from the standpoints of both theoretical and applied research. This was brought about, sometimes, by independent groups of researchers with different academic backgrounds who worked without the benefit of interaction between groups. Since the subject has a rich mathematical flavor, any attempt to promote dialogue between engineers emphasizing theory and mathematicians willing to gear their research towards applications, is, unquestionably, very beneficial. This book, then, gradually evolved out of the realization of the necessity of documenting the significant progress and novel directions of research in the arena of multidimensional systems theory, since the publication of my textbook, with the ultimate objective of alerting the reader about the existence of an apparently inexhaustible source of open problems, only some of which could be explicitly stated in this book.

The first chapter contains a description of very recent results in topics of fundamental importance in multidimensional systems theory. Perusal of this chapter is likely to convince the reader of the broadening scopes for applications of either some key theoretical results or limitations to such applications caused by problems, which still defy complete, satisfactory, solutions. Since, approximation theory is a very vast topic, only multivariate rational approximants of the Padé-type has been briefly surveyed in Chapter 2, because this type of rational approximation technique and its variants are popular in various problems of multidimensional systems theory. A rigorous analysis of a type of 2–D feedback system is provided in Chapter 3, where the advantages of incorporating the weakly causal feature in the design of stabilizing compensators is highlighted. Multidimensional feedback systems have been proposed in several applications, and the future role of feedback in the design of multidimensional structures should not be underestimated. In Chapter 4, the problems of existence and construction of feedback compensators for spatio-temporal systems, whose inputs and outputs are functions of a temporal and a spatial variable, are studied. The approach in Chapter 3 is a transform domain approach based on tools from algebra while that of Chapter 4 is based on techniques from the theory of linear systems with coefficients in a commutative ring and the criteria for stabilizability are specified in terms of a state-space representation (and also a transfer matrix representation). Chapter 5 introduces the reader to the modeling, analysis, and applications of linear, shift-variant multidimensional systems while Chapter 6 describes the theory of Gröbner bases which are known to have proven as well as potential applications in multidimensional systems theory problems. Chapter 7 considers conditions for the solution of a system of equations over the ring $C[z, w]$, motivated by the availability of some recent results in the mathematical literature and the applicability of the solution to the problem considered to one formulated in Chapter 3.

This book is aimed towards fulfilling the needs of those mathematicians who want to learn about current applications in an area requiring a wide variety of mathematical resources and also towards scientists, engineers, and researchers in industries and universities who want to keep abreast of latest developments in multidimensional systems theory, the directions along which this subject is expanding and some of the problems that are, currently, open to investigation. The book could also be, effectively, used in advanced seminars, selected continuing education courses and, possibly, as a supplement to a text adopted for courses in multidimensional systems theory or spatial and temporal signal processing.

The book could not be written without the cooperation of the contributors, some of whom, in addition to writing their respective portions, also made available some open problems for the benefit of the readers. Special thanks go to Dr. J. P. Guiver, who stimulated many interesting discussions and provided useful insights to some research problems during his stay, here, at the University of Pittsburgh. I have greatly benefited from interaction with colleagues at various universities, who made available to me many of their research results before those formally appeared in journals. The brief survey in Chapter 2 was partially influenced by the doctoral dissertation of A. Cuyt. The motivation and help provided by other colleagues is partly reflected in the contents of this book and I shall not attempt to acknowledge everyone here because it is very easy to, inadvertently, omit a name that would be expected to appear. I wish to express my sincere gratitude to the Air Force Office of Scientific Research, where I have been fortunate to work in Dr. Joseph Bram's program in the Directorate of Mathematical and Information Sciences, and to the National Science Foundation for continuing support of the research I have been conducting with my group of researchers. Some results of that research are included in this book. I thank Professor M. Hazewinkel for extending to me his kind invitation to contribute in his useful series and to the D. Reidel Publishing Company for their cooperation in the successful completion of this project.

348 Benedum Hall,   N. K. BOSE
University of Pittsburgh,
Pittsburgh, PA 15261, U.S.A.

Introducing the Contributors

N. K. Bose is a Professor of Electrical Engineering and Mathematics at the University of Pittsburgh. He is the author of the textbook on 'Applied Multidimensional Systems Theory', published by Van Nostrand Reinhold Co., NY, in 1982. He was elected to be a Fellow of IEEE for his contributions to multidimensional systems theory and circuits and systems education. His address is: 348 Benedum Hall, University of Pittsburgh, Pittsburgh, PA 15261, U.S.A.

B. Buchberger is a Professor of Mathematics at the University of Linz in Austria. He pioneered the development of the theory of Gröbner bases. His address is: Institut für Mathematik, Johannes Kepler Universität Linz, A-4040 Linz, Austria.

J. P. Guiver received his Ph.D. in mathematics at the University of Pittsburgh in 1982. The contributions associated with his name in this book are based on the research conducted at the University of Pittsburgh. He is currently with British Aerospace Dynamics Group, Bristol Division (FPC 213-Dept. 256), P.O. Box 5, Filton, Bristol BS12 7QW, England.

E. W. Kamen is a Professor of Electrical Engineering at the University of Florida, and Associate Director of the Center for Mathematical System Theory. His research interests center on the theory of linear infinite-dimensional systems (including systems with time delays), linear time-varying systems and linear multidimensional systems. His address is: Department of Electrical Engineering, University of Florida, Gainesville, FL 32611, U.S.A.

H. M. Valenzuela completed his Ph.D. in electrical engineering at the University of Pittsburgh in 1983. He is currently with the Dpto. Ing. Electrica, Universidad de Concepcion, Casilla 53-C, Concepcion, Chile.

Chapter 1

N. K. Bose

Trends in Multidimensional Systems Theory

1.1. INTRODUCTION

The theories of functions and polynomials of several complex and/or real
variables along with their numerous applications in several areas of
systems theory primarily concerned with the topics of multidimensional
digital filter, stability, stabilization and design, multivariate network
realizability theory, digital array processing in the general framework of
multidimensional signal processing techniques needed to process signals
carried by propagating wave phenomena, in addition to problems occurring in control theory concerned with 2–D state-space models, notions of
controllability, observability, minimality, feedback and pole-placement
as well as stabilization via output feedback provided the subject matter of
a Special Issue devoted to Multidimensional Systems [1.1]. Though the
vast majority of the papers in [1.1] were concerned with multidimensional
deterministic systems, a paper concerned with methods for inference
about models of random processes on multidimensional Euclidean space
from observed data was also included especially because the mathematical tools employed partially fitted those widely used in the Issue and
also because it was felt at that time that the future scopes for development
of multidimensional systems theory should not be restricted to deterministic systems only. In fact, broadly speaking, multidimensional
systems theory spans deterministic and statistical approaches to the
modeling, analysis and design of spatio-temporal continuous and discrete
systems. A collection of reprints geared particularly to the developments
in the area of multidimensional signal processing (deterministic as well as
statistical) till about 1977 has been compiled in [1.2]. As a result of
increasing research activity in the area and the realization of the necessity
to encourage interaction between computer scientists, engineers, and
mathematicians, a collection of reprints which characterized the significant developments that took place in the domain of multidimensional
systems theory till about 1979, was published in [1.3], where an intro-
ductory survey article and a list of open problems were also included. The
importance of documenting the scattered research results in an unified

1

N. K. Bose (ed.), Multidimensional Systems Theory, 1–40
© 1985 by D. Reidel Publishing Company.

form so that the theoretical fundamentals, even though based on relatively advanced and broad range of mathematical topics, could be presented to advanced graduate students and research scientists in a class-
room or seminar types of settings, the need was felt to select and expound
fundamental results of proven and potential significance in the area of
mathematical multidimensional systems theory. This was especially so
because though a sizeable number of books had appeared of an applied
nature, especially in the area of image processing, books devoted
exclusively to the more mathematical aspects that support such applications were non-existent at that time. One complete chapter in [1.4] was
devoted to multiparameter systems including two-dimensional filters,
distributed processes and statistical and probabilistic models for random
fields. More recently, a comprehensive account of the developments in
the theory of deterministic multidimensional systems along with an
exposition of the supporting mathematical tools required in selected
branches of study where such tools are used, was given in [1.5].

The objective here will be to document the progress in multidimensional systems theory since [1.5] was published so that the reader is
alerted to the flurry of activities generated by certain fundamental results,
some of which are finding applications in more than one area of research.
In the process it will become apparent that the topic of multidimensional
systems continues to provide challenging theoretical problems which
arise in the continuously increasing domains of applications. Efforts will
be made to relate briefly but succinctly past efforts, present status and
future trends so that the reader following his perusal of the contents of
this chapter will not only be able to appreciate better the contents of the
succeeding chapters but also will become cognizant of the resources
available to tackle the open problems, implicitly as well as explicitly
mentioned in this book. Though it is recognized that a fundamental
concept, result, or theoretical limitation, usually affects more than one
area of application and attempts to segment the range of applications to
distinct domains is sometimes futile, it is felt that when, for the sake of
clarity in exposition, it becomes necessary to consider one particular topic
in one particular section then the relevant links with other topics or
applications need to be cited and attended to. It is, therefore, hoped that
the topics selected for discussion in the following sections will represent
and reveal the existence of strong coupling between theoretical fundamentals and applications that fall under the umbrella of multidimensional
systems theory.

1.2. MULTIDIMENSIONAL SYSTEMS STABILITY

The concept of stability plays an important role in various areas that fall under the jurisdiction of multidimensional systems theory. Akin to the occurrence of several fundamental stability criteria in other fields of science (for example, stability of solutions of differential equations in nonlinear systems theory could be investigated under the definitions of Liapunov stability, asymptotic stability, conditional stability, orbital stability, stability under perturbation, etc.), the topic of multidimensional systems also uses the concept of stability in various context-dependent forms. An exposition of the state of this topic until about the end of 1981 is given in a recent book by Bose, [1.5], whose coverage of stability is geared towards linear shift-invariant multidimensional discrete systems, delay-differential systems, lumped-distributed and variable-parameter networks and to a lesser extent, stiff differential systems occuring in discussions of stability questions for numerical methods. Here, the important characteristics and developments of stability theory in these and related areas will be highlighted and attention will be directed to some questions that need to be satisfactorily resolved.

1.2.1. *Multidimensional Digital Filters*

The most commonly used criterion for assessing the stability of linear shift-invariant (LSI) multidimensional digital filters is the *bounded input/bounded output* (BIBO) criterion, which is well known to be equivalent to the requirement of absolute summability of the impulse response sequence characterizing the filter. A LSI n-dimensional recursive filter is also characterizable by a transfer function,

$$H(z_1, z_2, \ldots, z_n) = \frac{A(z_1, z_2, \ldots, z_n)}{B(z_1, z_2, \ldots, z_n)} \qquad (1.1)$$

where $H(z_1, z_2, \ldots, z_n)$ is viewed as a rational function of $z_1, z_2, \ldots, z_n$, $z_1^{-1}, z_2^{-1}, \ldots, z_n^{-1}$ (recursible filters include causal and weakly causal filters). For notational convenience, the z-transform $H(z_1, z_2, \ldots, z_n)$ of an n–D sequence, $\{h(k_1, k_2, \ldots, k_n)\}$, will be defined as a power series involving the superposition of the products of monomials of the type $z_1^{k_1} z_2^{k_2} \ldots z_n^{k_n}$ and the generic element, $h(k_1, k_2, \ldots, k_n)$, of the sequence. Physically, the indeterminates $z_1, z_2, \ldots, z_n$ are the respective delay variables along the spatial or temporal directions of sampling during the analog to digital conversion of a multidimensional spatio-temporal signal. In the case of first quadrant quarter-plane filters, $A(z_1, z_2, \ldots, z_n)$ and $B(z_1, z_2, \ldots, z_n)$ are polynomials. Since a ring isomorphism (see Chapters 3 and 5) maps a weakly causal filter onto a first quadrant quarter-plane filter, $A(z_1, z_2, \ldots, z_n)$, $B(z_1, z_2, \ldots, z_n)$, will be understood to be relatively prime *polynomials* in the delay variables $z_1, z_2, \ldots, z_n$ unless mentioned otherwise.

For a first quadrant quarter-plane digital filter, characterized by $H(z_1, z_2, \ldots, z_n)$, which is assumed to be holomorphic around the origin (this is assured by assuming $B(0, 0, \ldots, 0) \neq 0$), thereby permitting a Taylor series expansion,

$$H(z_1, z_2, \ldots, z_n) = \sum_{k_1 = 0}^{\infty} \cdots \sum_{k_n = 0}^{\infty}$$

$$h(k_1, \ldots, k_n) z_1^{k_1} \cdots z_n^{k_n}$$

$$(1.2)$$

the investigation into BIBO stability reduces to the determining of conditions under which the sum

$$\sum_{k_1 = 0}^{\infty} \cdots \sum_{k_n = 0}^{\infty} |h(k_1, \ldots, k_n)| \qquad (1.3)$$

converges.† Denoting by $\bar{U}^n$, U^n, and T^n the closed unit polydisc ($|z_i| \leq 1$, $i = 1, \ldots, n$), the open unit polydisc ($|z_i| < 1$, $i = 1, \ldots, n$), and the distinguished boundary ($|z_i| = 1$, $i = 1, \ldots, n$), it is well-known that convergence of (1.3) implies uniform convergence in $\bar{U}^n$ of (1.2), which in turn implies that $H(z_1, \ldots, z_n)$ is holomorphic in U^n and continuous on U^n. Also, if $H(z_1, \ldots, z_n)$ is holomorphic in a neighbourhood of $\bar{U}^n$, then (1.3) converges. In the $n = 1$ case, for a rational function, $H(z_1) = [A(z_1)]/[B(z_1)]$, it is simple to establish that (1.3) with $n = 1$ is absolutely summable if and only if the polynomial $B(z_1) \neq 0$, $|z_1| \leq 1$. This fact does not generalize in the $n > 1$ case and $B(z_1, \ldots, z_n) \neq 0$ in $\bar{U}^n$ is only a sufficient condition for BIBO stability of $H(z_1, \ldots, z_n)$ in (1.1). Though, in this context a set of necessary and sufficient conditions have not yet been obtained, some recent progress made towards the attainment of that goal is worth recording. The notations in [1.6] will be adopted. Let m_n be the Lebesgue measure divided by $(2\pi)^n$ that is carried with the compact

---

†Absolute convergence in (1.3) implies that $h(k_1, \ldots, k_n)$'s are uniformly bounded [1.66, p. 102], i.e. there exists a constant K such that, $|h(k_1, \ldots, k_n)| \leq K$, $k_1, \ldots, k_n = 0, 1, 2, \ldots$.

Abelian group (with componentwise multiplication as group operation) T^n, so that $m_n(T^n) = 1$. Also, for $0 < p < \infty$, let $H^p(U^n)$ be the class of all holomorphic functions, $F(z_1, \ldots, z_n)$ in U^n, for which,

$$\|F\|_p \triangleq \sup_{0 \leq r < 1} \left\{ \int_{T^n} |F(rz)|^p \, dm_n \right\}^{1/p} < \infty. \tag{1.4}$$

The following results, due to Dautov, will be stated and proved because of its unavailability in the western literature.

THEOREM 1.1. (*Dautov* [1.7]). *For the convergence of the summation in* (1.3), *it is sufficient that the function*

$$G(z_1, \ldots, z_n) \triangleq \frac{\partial^n[H(z_1, \ldots, z_n)z_1 \cdots z_n]}{\partial z_1 \partial z_2 \cdots \partial z_n}$$

belongs to $H^1(U^n)$.

Proof. By differentiation,

$$\frac{\partial^n\left[H(z_1, \ldots, z_n) \prod_{i=1}^{n} z_i \right]}{\partial z_1 \cdots \partial z_n} = \sum_{k_1 = 0}^{\infty} \cdots \sum_{k_n = 0}^{\infty} \tag{1.5}$$

$$(k_1 + 1) \cdots (k_n + 1)h(k_1, \ldots, k_n)z_1^{k_1} \cdots z_n^{k_n}.$$

Denote $g(k_1, \ldots, k_n) \triangleq (k_1 + 1) \cdots (k_n + 1)h(k_1, \ldots, k_n)$.
The Hardy–Littlewood inequality generalized to the n–D case yields

$$\sum_{k_1 = 0}^{\infty} \cdots \sum_{k_n = 0}^{\infty} \frac{|g(k_1, \ldots, k_n)|}{(k_1 + 1) \cdots (k_n + 1)} \leq \pi^n \|G\|_1. \tag{1.6}$$

From (1.5) and (1.6), the proof of the theorem is easily completed.

The following definition must be given before stating the next result of Dautov, which is useful to check whether a function belongs to $H^1(U^n)$.

DEFINITION 1.1. If $F(z_1, \ldots, z_n)$ is any function in U^n, we define $\hat{F}(z_1, \ldots, z_n)$ by (note that $\mathbf{z} = (z_1, \ldots, z_n)$),

$$\hat{F}(\mathbf{z}) = \lim_{r \to 1} F(r\mathbf{z})$$

at every $\mathbf{z} \epsilon T^n$ at which this radial limit exists.

Let $L^1(T^n)$ denote the space of all functions integrable with respect to the measure m_n. Then, the following result was proved by Dautov.

THEOREM 1.2. [1.7]: If $F(\mathbf{z})$ is a rational function (in reduced form) whose denominator is zero only in a finite number of points belonging to T^n and $\hat{F}(\mathbf{z}) \epsilon L^1(T^n)$, then $F(\mathbf{z}) \epsilon H^1(U^n)$.

Dautov, subsequently, arrived at results on stability for special classes of rational functions, whose denominator polynomial has a finite number of zeros on T^n. These results are based on the definition for the order of a zero introduced next.

DEFINITION 1.2. For the sake of brevity, denote $|k| = k_1 + \ldots + k_n$, where $k_i > 0$, $\forall\, i$ and $k = (k_1, k_2, \ldots, k_n)$, so that the power series in (1.2), rewritten as (to avoid cluttering of notation let $\mathbf{z}^k$ denote $z_1^{k_1} \ldots z_n^{k_n}$),

$$H(\mathbf{z}) = \Sigma \ldots \Sigma h(k)\mathbf{z}^k$$

is in some polydisc D^n centered at the origin in C^n ($C^n \triangleq C \times C \ldots \times C$, is the Cartesian product of n copies of the complex field C). For $s = 0, 1, 2, \ldots$ let $F_s(\mathbf{z})$ be the sum of those terms, $h(k)\mathbf{z}^k$, for which $|k| = s$, so that

$$H(\mathbf{z}) = \sum_{s=0}^{\infty} F_s(\mathbf{z}).$$

If $H(\mathbf{z})$ is not identically zero in D^n, then there is a smallest $s = 0_o$ such that F_{0_o} is not the zero polynomial. This 0_o is the order of the zero which $H(\mathbf{z})$ has at the origin. Furthermore, if $H(\mathbf{z})$ is holomorphic in a neighbourhood of $\mathbf{z}^{(0)}$, then the order of the zero of $H(\mathbf{z})$ at $\mathbf{z}^{(0)}$ is the order of the zero of $H(\mathbf{z} + \mathbf{z}^{(0)})$ at $z = 0$.

Next, consider an irreducible rational function,

$$H(\mathbf{z}) = \frac{A(\mathbf{z})}{1 - b_1 z_1 - \ldots - b_n z_n} \tag{1.7}$$

where the b_k's belong to the complex field C. It is elementary to verify that the denominator polynomial of $H(\mathbf{z})$ in (1.7) is nonzero in $\bar{U}^n$ if and only if $\sum_{k=1}^{n} |b_k| < 1$. In case $\sum_{k=1}^{n} |b_k| = 1$, this denominator polynomial is zero in $\bar{U}^n$ if and only if

$$z_k = \frac{b_k^*}{|b_k|}, \qquad k = 1, 2, \ldots, n. \tag{1.8}$$

The zero, then, is at

$$\mathbf{z}^{(0)} = \left(\frac{b_1^*}{|b_1|}, \ldots, \frac{b_n^*}{|b_n|} \right). \tag{1.9}$$

Note that $\mathbf{z}^{(0)}$ is on T^n.

THEOREM 1.3. [1.7] $H(\mathbf{z})$ in (1.7) *characterizes a BIBO stable system even when it has a nonessential singularity at* $\mathbf{z}^{(0)}$ *on* T^n *(due to the constraint,*

$\sum_{k=1}^{n} |b_k| = 1$) *if for polynomial* $A(\mathbf{z})$, *the zero order at point* $\mathbf{z}^{(0)}$ *satisfies*

$$0_{\mathbf{z}^{(0)}}(A) \geq 2(n + 1).$$

Proof. $B(\mathbf{z}) = 1 - b_1 z_1 - \ldots - b_n z^n$.

For any complex variable w,

$$|w| \leq 1 \leftrightarrow 2 \operatorname{Re}(1 - w) \geq |1 - w|^2. \tag{1.10}$$

In (1.9), $z_k^{(0)} = b_k^* |b_k|^{-1}$. Define,

$$w_k \triangleq (z_k^{(0)})^{-1} z_k = |b_k| [b_k^*]^{-1} z_k, \qquad k = 1, \ldots, n. \tag{1.11}$$

Using the constraint, $\sum_{k=1}^{n} |b_k| = 1$ and (1.11),

$$|B(z)| \geq \operatorname{Re} B(z) \geq \operatorname{Re} \sum_{k=1}^{n} |b_k|(1 - w_k). \tag{1.12}$$

Since $|z_k| \leq 1$ implies that $|w_k| \leq 1$, $\mathbf{z} \epsilon \bar{U}^n$ implies that $\mathbf{w} \epsilon \bar{U}^n$. Substituting (1.10) in (1.12).

$$|B(\mathbf{z})| \geq \tfrac{1}{2} \sum_{k=1}^{n} |b_k| |1 - w_k|^2 \geq$$

$$\tfrac{1}{2} \min \{|b_1|, \ldots, |b_n|\} \sum_{k=1}^{n} |1 - w_k|^2. \tag{1.13}$$

Substituting (1.11) in (1.13),

$$|B(\mathbf{z})| \geq K|\mathbf{z} - \mathbf{z}^{(0)}|^2, \qquad \mathbf{z} \epsilon \bar{U}^n \tag{1.14}$$

where,

$$K \triangleq \tfrac{1}{2} \min \{|b_1|, \ldots, |b_n|\} \quad \text{and}$$

$$|\mathbf{z} - \mathbf{z}^{(0)}|^2 = \sum_{k=1}^{n} |z_k - z_k^{(0)}|^2$$

is the Euclidean norm of $(\mathbf{z} - \mathbf{z}^{(0)})$.

By successive differentiation, it is possible to verify that

$$\frac{\partial^n H(\mathbf{z})\, z_1 \cdots z_n}{\partial z_1 \cdots \partial z_n} = \frac{A_1(\mathbf{z})}{[B(\mathbf{z})]^{n+1}} \tag{1.15}$$

where $A_1(\mathbf{z})$ is a polynomial whose zero order is not less than the zero order of $A(\mathbf{z})$. From the condition in the theorem, it then follows that the zero order of $A_1(\mathbf{z})$ at $\mathbf{z} = \mathbf{z}^{(0)}$ is given by,

$$0_{\mathbf{z}^{(0)}}(A_1) \geq 2(n + 1).$$

Therefore,

$$|A_1(\mathbf{z})| \leq K_1|\mathbf{z} - \mathbf{z}^{(0)}|^{2(n+1)}, \qquad \mathbf{z} \epsilon \bar{U}^n \tag{1.16}$$

where K_1 is a positive constant. Therefore, from (1.14), (1.15), and (1.16),

$$\frac{\partial^n H(\mathbf{z})\, z_1 \cdots z_n}{\partial z_1 \cdots \partial z_n} \leq \frac{K_1}{K^{n+1}}, \qquad \mathbf{z} \epsilon \bar{U}^n, \qquad \mathbf{z} \neq \mathbf{z}^{(0)}. \tag{1.17}$$

Subsequently, using Theorems 1.1 and 1.2, Dautov mentioned the feasibility for demonstrating validity of this Theorem by generalizing techniques employed to tackle special cases of the denominator polynomial in (1.7).

For an arbitrary rational function $H(\mathbf{z})$, the problem of determining conditions for BIBO stability in the presence of a finite number of nonessential singularities of the second kind on T^n is complex and, at present, no general solution is available. Dautov [1.8] clarified the problem in the 2–D case by showing that for a certain class of denominator polynomials $B(\mathbf{z})$, $H(\mathbf{z}) = A(\mathbf{z})/B(\mathbf{z})$ is BIBO stable if and only if it can be continuously extended to $\bar{U}^2$ from U^2; furthermore, he conjectured that this is true for any $B(\mathbf{z})$ whose only zeros in $\bar{U}^2$ are on T^2 and these are also zeros of $A(\mathbf{z})$ (note that in the 2–D case the finiteness condition for the number of common zeros on T^n is automatically satisfied). Of course, $H(\mathbf{z})$ can never be extended *analytically* to $\bar{U}^n$ when there is a nonessential singularity of the second kind at $z = z^{(0)} \in T^n$ since in any open neighbourhood of $z = z^{(0)}$, $H(z)$ will always be unbounded and therefore Riemann's analytic continuation theorem will not apply. However, if $H(z)$ extends *continuously* to $\bar{U}^n$ one can take limits from any direction from within $\bar{U}^n$ to achieve the same result. The example given next shows a rational function $H(z)$ which is analytic in U^2, has a non-essential singularity of the second kind at $z = (1,1)$ and does not extend on $\bar{U}^2$ to a continuous function.

EXAMPLE 1.1. Let,

$$H(z_1, z_2) = \frac{(z_1 - 1)^n + (z_1 - 1)(z_2 - 1) + (z_2 - 1)^n}{(z_1 + z_2 - 2)^2},$$

$$n \gg 1.$$

For $(z_1, z_2) = (e^{j\theta}, e^{j\phi}) \in T^2$,

$$H(e^{j\theta}, e^{j\phi}) = \frac{(e^{j\theta} - 1)^n + (e^{j\theta} - 1)(e^{j\phi} - 1) + (e^{j\phi} - 1)^n}{(e^{j\theta} - 1 + e^{j\phi} - 1)^2}.$$

Then, the limit as $\theta \to 0$ along the line $\phi = 0$ is 0, but the limit as $\theta \to 0$ along the line $\theta = \phi$ is $\frac{1}{4}$. (This example is due to Dr. J. Murray.)

The famous counterexample of Goodman [1.9] also exhibits this property. Through a series of neatly constructed examples, Goodman, in a prize-winning paper [1.9], was the first to point out difficulties in the prevailing concept of BIBO stability for multidimensional filters. In addition to other results, he showed that

$$H(z_1, z_2) = \frac{(1 - z_1)^m (1 - z_2)^n}{2 - z_1 - z_2}, \qquad m \geq 0, \qquad n \geq 0 \quad (1.18)$$

is BIBO unstable when $m = n = 1$, and BIBO stable when $m = n = 8$. For the case when $m = n = 1$ it is clear that $H(z_1, z_2)$ in (1.18) does not extend on $\bar{U}^2$ to a continuous function, since

$$\lim_{z_1 \to 1} H(z_1, z_1) = \lim_{z_1 \to 1} \frac{(1 - z_1)^2}{2(1 - z_1)} = 0$$

while

$$\lim_{x \to 0} H(1 - x^2 + jx, 1 - x^2 - jx) = \lim_{x \to 0} \frac{x^4 + x^2}{x^2} = 1.$$

Therefore, the validity of Dautov's conjecture (see also 'Open Problems', Problem #1 where general conjectures have been made), will enable one to arrive at Goodman's conclusions concerning specializations of (1.18) in a relatively simple manner. Furthermore, it follows from Dautov's results [1.8], that $H(z_1, z_2)$ is BIBO stable when $m = n$ for all positive powers of the numerator in (1.18) except when $m = n = 1$. In fact, it is easy to infer from [1.8] that $H(z_1, z_2)$ in (1.18) is BIBO stable except when the 2-tuple (m, n), takes values $(0, 0)$, $(0, 1)$, $(1, 0)$, $(1, 1)$, $(2, 0)$, and $(0, 2)$.

Though Dautov's conjecture on 2–D filter BIBO stability in the presence of nonessential singularities of the second kind on T^2 is still unresolved at least in the open literature, a sufficient condition and a necessary condition expressed in terms of tangents to the algebraic curve at a zero of $B(z_1, z_2)$ on T^2 have been advanced by Alexander and Woods [1.10]. This necessary condition uses the fact that when $n = 2$ in (1.1) the number of nonessential singularities of the second kind is finite and isolated, while in the $n > 2$ case the complexity of the problem increases considerably since the locus of those singularities is of real dimension $2n$–4 in a space of real dimension $2n$ and these singularities cannot disconnect the space. The sufficient condition for BIBO stability, as given by Alexander and Woods, uses a result by Zak [1.11] involving an extension to two variables of a well-known theorem on the absolute convergence of the Fourier series expansion of a periodic function of bounded variation and belonging to class Lipschitz α. Though the problem of the resolution of nonessential singularities of the second kind in the context of BIBO stability of multidimensional filters remains to be satisfactorily tackled, the tests for the absence of zeros of a multivariate polynomial in a unit polydisc,

$$B(z_1, z_2, \ldots, z_n) \neq 0, \qquad \mathbf{z} \triangleq (z_1, z_2, \ldots, z_n) \epsilon \bar{U}^n \qquad (1.19)$$

can be implemented in various forms, proceeding from criteria resulting from Rudin's theorem, which itself is provable by using simple one-variable arguments exclusively, as done by Delsarte–Genin–Kamp [1.12]. Note that $B(0, 0, \ldots, 0) \neq 0$, the condition necessary to permit the power series expansion in (1.2) of the rational function $H(\mathbf{z})$, allows the test in (1.19) to be replaced by the test in (1.20), where m_1 is the degree in z_1 of polynomial $B(z_1, \ldots, z_n)$

$$z_1^{m_1} B(z_1^{-1}, z_2, \ldots, z_m) \neq 0,$$

$$z_1 \notin U^1, \qquad (z_2, \ldots, z_{n-1}) \in \bar{U}^{n-1}. \tag{1.20}$$

The various algebraic tests to test for the conditions in (1.19) and (1.20) are given in [1.5], [1.13], and [1.14].

1.2.2. Multivariate Networks

Various types of electrical networks like lumped-distributed and variable parameter networks can be studied within the setting of multidimensional systems theory. The various definitions for multivariate Hurwitz polynomials and its ramifications have been discussed by Bose [1.5] in his exposition of the realizability theory for multivariate rational functions in the complex variables $\mathbf{p} \triangleq (p_1, p_2, \ldots, p_n)$. Recently, Fettweis [1.15] has introduced a particular class of Hurwitz polynomials, referred to as *principal Hurwitz* polynomials or *scattering Hurwitz* polynomials in view of their central role in the scattering transfer matrix description of multivariate lossless two ports encountered in the synthesis of reference filters as a prelude to the derivation of wave digital filter structures. For the sake of generality, complex coefficients will be permitted in a polynomial.

DEFINITION 1.3. A polynomial, $g(\mathbf{p})$ will be called a scattering Hurwitz polynomial or, equivalently, a principal Hurtwitz polynomial, if it satisfies the following properties:

(a) $g(\mathbf{p}) \neq 0$, Re $\mathbf{p} > 0$, i.e. the open right-half polydomain and

(b) $g(\mathbf{p})$ is relatively prime to the paraconjugate polynomial, $g_*(\mathbf{p}) \triangleq g^*(-\mathbf{p}^*) = g^*(-p_1^*, -p_2^*, \ldots, -p_n^*)$, where the star superscript represents the operation of complex conjugation. (Note that when the polynomial coefficients are real, $g_*(\mathbf{p}) = g(-\mathbf{p}) = g(-p_1, -p_2, \ldots, -p_n)$).

Subsequently, Fettweis showed the equivalence of Definition 1.3 to Definition 1.4 given next in a slightly specialized but somewhat simpler form than originally proposed.

DEFINITION 1.4. A polynomial, $g(\mathbf{p})$, is a scattering Hurwitz polynomial if and only if

(a) $g_*(\mathbf{p})$ and $g(\mathbf{p})$ are relatively prime and

(b) $\dfrac{|g_*(\mathbf{p})|}{|g(\mathbf{p})|} \leq 1$, Re $\mathbf{p} > 0$.

The key to the proof of the equivalence of the preceding two definitions is the development (and use) of the maximum-modulus principle to situations where nonessential singularities of the second kind are allowed to occur on the boundary of the open polydomain, Re $\mathbf{p} > 0$, for the rational function $[g_*(\mathbf{p})]/[g(\mathbf{p})]$. An interesting outcome of the introduction of the previous two definitions is the multivariate counterpart of a well-known result in passive network synthesis. Define the para-even and para-odd parts, respectively, of $g(\mathbf{p})$.

$$g_e(\mathbf{p}) = \frac{[g(\mathbf{p}) + g_*(\mathbf{p})]}{2}, \qquad g_o(\mathbf{p}) = \frac{[g(\mathbf{p}) - g_*(\mathbf{p})]}{2}. \qquad (1.21)$$

Then, the following result holds. Let $g(\mathbf{p})$ have real coefficients.

FACT 1.1. If $g(\mathbf{p})$ is a scattering Hurwitz polynomial of degree ≥ 1, then the quotient of $g_e(\mathbf{p})$ and $g_o(\mathbf{p})$ is a nontrivial (that is not zero or an imaginary constant) multivariate reactance function in irreducible form. Vice versa, if the quotient of two relatively prime polynomials, $g_e(\mathbf{p})$ and $g_o(\mathbf{p})$, is a nontrivial reactance function then their sum is a scattering Hurwitz polynomial of degree ≥ 1.

It is mentioned that other definitions used with regard to multivariate Hurwitz polynomials include the imposing of restrictions, $g(\mathbf{p}) \neq 0$, Re $\mathbf{p} > 0$ ($g(\mathbf{p})$ is then, referred to as *widest sense Hurwitz*) and $g(\mathbf{p}) \neq 0$, Re $\mathbf{p} \geq 0$ ($g(\mathbf{p})$ is, then, referred to as *strictest sense* Hurwitz), $g(\mathbf{p}) \neq 0$, Re $p_i > 0$, Re $p_j = 0$ when $j \neq i$ for $i = 1, 2, \ldots, n$, $j = 1, 2, \ldots, n$, $g(\mathbf{p}) \neq 0$, Re $p_i = 0$, Re $p_j > 0$ when $j \neq i$ for $i = 1, 2, \ldots, n$, $j = 1, 2, \ldots, n$ ($g(\mathbf{p})$, in the previous two cases, when $n = 2$, has been called *narrow sense* Hurwitz) and various other obvious ramifications. Widest sense Hurwitz polynomials have been recently considered by Gregor [1.16]. It should be noted that the test for the scattering Hurwitz property requires the test for a polynomial to be widest sense Hurwitz. While algebraic tests exist to determine whether or not a polynomial is strictest sense Hurwitz, efficient tests for establishing whether or not a polynomial is devoid of zeros in the open right-half polydomain are now being developed. A strictest sense Hurwitz polynomial is obtainable from a polynomial $B(z_1, z_2, \ldots, z_n)$ devoid of zeros in $\bar{U}^n$ after bilinearly transforming each variable z_i,

$$z_i \to \frac{1 - p_i}{1 + p_i}, \qquad i = 1, 2, \ldots, n \qquad (1.22)$$

and then multiplying the resulting rational function by the denominator polynomial in $p_1, p_2, \ldots, p_n$. However, for a strictest sense Hurwitz polynomial to be 'bilinearly transformable' to a polynomial devoid of zeros in $\bar{U}^n$ requires some additional restrictions. These restrictions have been treated by Bose [1.5] with pertinent references.

In this section, it is pointed out that microwave circuits which can be modeled by lumped network elements. and commensurate or non-commensurate (at least, for theoretical reasons) transmission lines can be analyzed and synthesized via transform techniques utilizing rational functions in several complex variables. In effect, an univariate trans-cendental function characterization is transformed into a multivariate rational function description of the system. This idea permeates an approach of analysis for delay-differential systems, to be described in the next section. A valid question in the stability analysis of lumped-distributed networks has been posed as a problem [1.5].

PROBLEM. Prove or disprove the following statement: The $(n + 1)$-variate polynomial $G(p_1, p_2, \ldots, p_{n+1})$ has no zeros in the polydomain Re $p_i > 0, i = 1, \ldots, n + 1$ (i.e. it is widest sense Hurwitz) if and only if the function $G(p_1, \tanh \alpha_1 p_1, \ldots, \tanh \alpha_n p_{n+1})$ has no zeros in Re $p_1 > 0$ for all $\alpha_i > 0, i = 1, 2, \ldots, n$. It may be assumed that $G(p_1, p_2, \ldots, p_{n+1})$ is devoid of polynomial factors of the form $\Pi_{k=2}^{n+1} (p_k - 1)^{m_k}$ or more specifically, albeit restrictively, $G(p_1, 1, 1, \ldots, 1) \not\equiv 0$.

A proof of the statement in the preceding problem has been given by Delsarte, Genin, Kamp [1.17], [1.18]. It is noted that the stability criterion under discussion relates the zero exclusion of a multivariate polynomial from a specified polydomain to the stability problem associ-ated with passive lumped-distributed networks, independent of delay. For any specified fixed delay, the assessment of stability becomes a more difficult problem, and it is not clear whether multivariate techniques will offer any advantage over those techniques that are currently available to handle zero exclusion problems for classes of univariate transcendental functions.

1.2.3. Delay-Differential Systems

Given a delay differential system of the retarded typed with delays equal to integer multiples of a fixed delay $h \geq 0$, it is well-known that the system is *asymptotically stable independent of delay* if and only if the characteristic function, $Q(p, e^{-hp})$, which may be viewed as a polynomial in p and e^{-hp}, satisfies the condition,

$$Q(p, e^{-hp}) \neq 0, \qquad \text{Re } p \geq 0. \tag{1.23}$$

Kamen [1.26] showed that after introducing an additional complex variable z to replace e^{-hp}, the bivariate polynomial,

$$Q_1(p, z) \triangleq Q(p, e^{-hp})|_{e^{-hp} \to z}$$

displays the following property.

FACT 1.2. $Q_1(p, z) \neq 0$, Re $p \geq 0$, $|z| = 1$ if and only if
   (a) (1.23) holds for all real $h \geq 0$ and
   (b) $Q(p_0, e^{-hp_0}) \neq 0$, $p_0 = j\omega/h$, $\omega \in [0, 2\pi]$, $\lim h \to \infty$.
For delay-differential equations of a more general type (including neutral type) with commensurate and incommensurate delays, some complications may arise. Consider a delay-differential equation of the generic type,

$$L(y) = \sum_{k=0}^{m_1} \sum_{r=0}^{m_2} a_{kr} \frac{d^k}{dt^k} y(t - h_r) = 0 \tag{1.24}$$

where a_{kr} and $h_r (k = 0, 1, \ldots, m_1; r = 0, 1, \ldots, m_2)$ are constants; furthermore,

$$0 = h_0 < h_1 < \ldots < h_{m_2} \tag{1.25a}$$

and

$$\begin{aligned} a_{km_2} &\neq 0 \quad \text{for some} \quad k, \, a_{m_10} \neq 0, \\ a_{m_1r} &\neq 0 \quad \text{for some} \quad r \neq 0. \end{aligned} \tag{1.25b}$$

Brumley [1.19] showed that even if every zero, $p_1 = p_1^{(0)}$ of the characteristic equation,

$$Q(p_1, e^{-p_1}) = \sum_{k=0}^{m_1} \sum_{r=0}^{m_2} a_{kr} p_1^k e^{-h_r p_1} \tag{1.26}$$

satisfies Re $p_1^{(0)} < 0$ and the h_r's are commensurable, it is possible for (1.24) to have unbounded solutions unless an additional polynomial condition is satisfied. The quasipolynomial in (1.26) is an entire analytic function in the complex variable p_1, possessing an infinite number of zeros, the only limit point of which is infinity. Even with the spectra of

(1.26) in the left half plane, it is possible that in some isolated cases due to the presence of roots which approach the imaginary axis at infinity, one may be able to construct a solution of (1.24), subject to appropriate initial conditions, which increases without limit for some sequence of values of the argument. Gromova [1.20] called such a distribution of roots of (1.26), for which there exists sequences of roots which approach the imaginary axis at infinity, the asymptotically critical case. When the h_r's in (1.25a) are incommensurable, the technique of Brumley in [1.19] does not apply.

Setting aside the pathological or isolated cases referred to above, the situations when the asymptotic stability of neutral differential equations is equivalent to their spectra lying in the left-half plane are of very great interest. Guiver and Bose [1.21] have considered several equivalences in the bivariate polynomial formulation of the test for asymptotic stability independent of delay for delay-differential equations of the neutral type.

1.2.4. *Stiff Differential Systems*

Genin [1.22] associated a bivariate canonical polynomial to a multistep-multiderivative formula used for integrating stiff differential equations. Specifically, consider a first order differential equation,

$$Dy = \frac{dy}{dt} = f_1(y, t). \tag{1.27}$$

An approximate solution x of (1.27) can be obtained through the use of the following linear multistep-multiderivative (LMSD) formula

$$\sum_{i=0}^{n} \sum_{j=0}^{k} (-1)^i a_{ij} h^i D^i x_{t-j} = 0, \tag{1.28}$$

$$t = k, k + 1, \ldots$$

initialized by a set of k starting values, $(x_0, x_1, \ldots, x_{k-1})$ for fixed integers n and k, real constants a_{ij} and $D^0 x_{t-j} = x_{t-j}$ is the computed value of y at time t_{t-j}, $D^i x_{t-j} \triangleq \dot{f_i}(x_{t-j}, t_{t-j})$ with $D^i y = f_i(y, t) \equiv (\partial f_{i-1}/\partial y) f_1 + (\partial f_{i-1}/\partial t)$ and h is the step size. The associated bivariate canonical polynomial is:

$$H(p_1, p_2) = \sum_{k=0}^{n} b_k(p_1) p_2^k$$

where

$$b_i(p_1) = (p_1 - 1)^k \sum_{j=0}^{k} a_{ij} \left(\frac{p_1 + 1}{p_1 - 1} \right)^{k-j}.$$

The (k, n) method based on (1.28) is *weakly stable* provided the polynomial $b_0(p_1) = H(p_1, 0)$ has exact degree $k - 1$, does not vanish in Re $p_1 > 0$ and has at most zeros of multiplicity 1 on Re $p_1 = 0$. When $b_0(p_1) \neq 0$ on Re $p_1 = 0$, the (k, n) method becomes *strongly stable*. A weakly stable (k, n) integration formula is said to have the error order ν and the corresponding error constant $K_{\nu+1}$ if its associated canonical polynomial satisfies

$$\frac{1}{p_1^k} H\left(p_1, -\log \frac{p_1 + 1}{p_1 - 1} \right) \sim K_{\nu+1} \left(\frac{2}{p_1} \right)^{\nu+1} \quad \text{for} \quad p_1 \to \infty.$$

The above remains satisfied if single variable factors, when present, of $H(p_1, p_2)$ are factored out. Under these conditions, Genin [1.22] showed that the integration formula (1.28) is A-stable if and only if $H(p_1, p_2)$ is a bivariate Hurwitz polynomial in the narrow sense. According to the Daniel–Moore conjecture, the maximum error order achievable by an n-derivative formula equals $2n$ and the maximum value of the corresponding error constant is

$$\frac{(n!)^2}{(2n)! \, (2n + 1)!}.$$

Recently, Delsarte–Genin–Kamp [1.23] proved the validity of the Daniel–Moore conjecture when $n = 2$ (others proved the validity in general via different approaches) using certain interesting properties of the bivariate canonical polynomial associated with a $(k, 2)$ integration formula. These properties are expressible in terms of the *positive realness* of certain univariate rational functions. The property of positive realness is known to form the nucleus of the topic of passive network synthesis. The possibility for relating bivariate positive functions to stability questions for numerical methods has been raised and remains largely unexplored except for a brief mention of Dahlquist [1.24] in reference to a study of A-stability 'in an implicit alternating direction scheme for hyperbolic and parabolic equations'.

1.2.5. *Multipass Processes*

A large variety of operations in the coal cutting, metal shaping, and automatic agricultural ploughing can be modeled as multipass processes, which are characterized by repetitive operations and interaction between the state and/or output variables generated during successive cycles of operation. Edward and Owens [1.25] have noted that the notion of stability along each individual cycle of operation, referred to as a *pass*, coincides with the notion of BIBO stability in the theory of 2–D systems. Consequently, the well developed stability tests in the area of 2–D digital filters can be directly applied to the stability analysis of multipass processes that may be characterized by linear shift-invariant discrete state-space models.

The status of stability theory for multidimensional systems has been reported. The broad scope of applications of fundamental tests is highlighted. The occurrence and utilization of definitions on various forms of stability are underscored. Since LSI multidimensional systems have been the prime target of research over the last decade, transform techniques have been primarily used. With attention being directed more and more to spatially-varying and non-linear multidimensional systems, the need for new tools and approaches to stability analysis is being felt.

1.3. MULTIVARIATE REALIZATION THEORY

In response to the scopes for applications of multidimensional systems theory, there has been a substantial advance in the status of realization theory. Generally speaking, the approaches adopted in the development of such a theory may be broadly classified under either state-space or transform domain (or polynomial matrix) categories. The state-space approach towards realization has been popular in estimation, identification and control-theoretic applications, while the techniques based on transform methods occur more within the settings of network theory, digital filtering, and related fields of study. However, any attempt to restrict a particular approach to selected areas of application may be unwise and futile.

1.3.1. *State-space Realization Theory*

Linear dynamical systems, discrete or continuous, characterized respectively by

$$\dot{x}(t) = Fx(t) + Gw(t), \qquad y(t) = Hx(t) + Jw(t) \qquad (1.29a)$$

or

$$x(k + 1) = Fx(k) + Gw(k), \qquad y(k) = Hx(k) + Jw(k)$$
$$(1.29b)$$

where F, G, H, J are constant matrices and x, u, y are the state, input and output vectors have been widely studied for over two decades. Since the presence or absence of J in (1.29a) or (1.29b) does not complicate in any way the treatment of the realization problem from the mathematical standpoint, for the sake of brevity in exposition, it will, henceforth be considered absent. Then, in the discrete case, the realization problem involves the determination of $\{F, G, H\}$, if they exist, such that for a specified impulse response sequence, $\{T_1, T_2, \ldots\}$, the following holds.

$$T_k = HF^{k-1}G, \qquad k = 1, 2, \ldots. \qquad (1.30)$$

When the elements of T_k and likewise of F, G, H are restricted to belong to any specified field, finite or infinite, procedures to obtain a realization $\{F, G, H\}$, the notions of minimality, reachability, observability and equivalence or isomorphism between two realizations have been thoroughly documented. The need for analysis and synthesis of a wider class of systems than those characterized by (1.29a) or (1.29b) necessitated the initiation of research into the realization theory of systems over rings. The first question that occurs when lifting the restrictions imposed by a field, is under what circumstances the finiteness criterion imposed on the sequence $\{T_1, T_2, \ldots\}$ (for realization to be possible, the infinite block-Hankel matrix generated with $T_1, T_2, \ldots$ in its first block row must have a finite rank) continues to hold. Via imposition of a very mild restriction on a commutative ring R, Rouchaleau, Wyman, and Kalman [1.27]. were able to prove the following result based on the definition that a ring R is Noetherian if every ideal is finitely generated. R is assumed to be an integral domain permitting the consideration of its quotient field K.

FACT 1.3. Let R be a Noetherian integral domain, K its quotient field, T_1, $T_2, \ldots$ an input/output sequence over R. This has a realization over R if and only if it is realizable over K.

The next step in classical linear system theory relates the notions of canonical (i.e. both reachable and observable) and minimal realizations under the imposition of equivalence for the dimensions of realizations;

this dimension equals the rank of the infinite block Hankel matrix formed from $T_1, T_2, \ldots$, as referred to previously. Rouchaleau and Sontag[1.28] defined a realization of an input/output sequence over R to be *absolutely minimal* if and only if its dimension is the same as that of a minimal (canonical in this case) realization of the specified input/output sequence over the quotient field K and proved the following general result.

FACT 1.4. The canonical realization of every input/output sequence over a Noetherian domain R is absolutely minimal if and only if R is a principal ideal domain, i.e. R is Noetherian and every pair of elements r_1, r_2 in R has a greatest common divisor which can be expressed as a linear combination, $r_3 r_1 + r_4 r_2$ where r_3, r_4 also belong to R.

The existence of absolutely minimal realizations which, however, may not be canonical can be guaranteed over rings more general than principal ideal domains. In multidimensional systems theory, these general rings include polynomials in not more than two indeterminates, having coefficients in a field. Sontag[1.29] also gave a lattice characterization for the class of minimal-rank realizations over a commutative Noetherian integral domain.

It has been noticed that unlike in the case of systems over fields, where a system is canonical (reachable and observable) if and only if it is minimal, for systems over rings a reachable and observable system is minimal but not, necessarily, vice-versa. Sontag [1.30] studied the observability properties of realizations of linear response maps over commutative rings and gave a characterization for those maps which admit realizations which are simultaneously reachable and observable in a strong sense. To do this (and also to tackle the problem of regulation of linear systems over commutative rings), he introduced the concept of split map described below for the case when R is a Noetherian integral domain, as this restriction is usually met by rings encountered in systems theory and the description also provides an algebraic criterion to test for the split property.

FACT 1.5. Let R be a Noetherian integral domain and K be its quotient field. Let $\{T_1, T_2, \ldots\}$ be an input/output matrix sequence over R and let n be the dimension of a canonical realization of this sequence over K. Let H_n be the block Hankel matrix whose first block row is $T_1, T_2, \ldots T_n$ and the last block column is formed from $T_n, T_{n+1}, \ldots, T_{2n-1}$ arranged sequentially from top to bottom. Then the input/output map is split if and only if the ideal generated by all $(n \times n)$ minors of H_n is R (or in other words the greatest common divisors of these minors is a unit, when R is a Bezout domain).

A survey of early developments in the state-space realization theory over commutative rings outlined above is given in [1.31]. In multidimensional systems theory Eising [1.32] viewed two-dimensional causal and weakly causal digital filter transfer matrices as linear systems over the rings of proper rational functions and stable proper rational functions, as a prelude to the development of a state-space realization model, for which the state can be recursively computed so long as the support of the impulse response sequence belongs to a causality cone [Chapter 3]. Other state-space models encountered in the realization of 2–D systems have been discussed by Bose [1.5]. State-space modeling of 3–D systems has been recently considered by Tzafestas and Pimenides [1.33], [1.34].

It is relevant to point out that some research into the realization theory over a noncommutative ring has also been conducted. The distinguishing fact here is that the Cayley-Hamilton theorem which provides the finiteness condition in the realization theory over a field or a commutative ring, fails to hold. Fliess [1.35] began the study on realization of rational power series in several non-commuting variables, after recognizing their relevance in the analysis of bilinear systems and possibly, in a larger class of nonlinear systems. Sontag [1.36] obtained some purely algebraic extensions of results in the theory of linear dynamical systems for the case when the coefficient ring is arbitrary. Fornasini [1.37] studied the possibility of using rational noncommutative power series to realize spatial filters. He presented an extension of Ho's algorithm in classical linear system theory, tackled the partial representation problem in which one is attempting to obtain a recursive model for the coefficients of a series on the basis of incomplete data, and finally considered the problem of generating all minimal realizations of a specified filter.

1.3.2. *Transform Domain Realization Theory*

The polynomial matrix approach to linear systems, initiated by Rosenbrock, is very useful in the study of realization theory and related problems of dynamic compensation, stabilization, output regulation in the presence of disturbances, input tracking, etc. It is well known that for a realizable input–output map over a field, the transfer matrix $T(p)$, whose elements are rational functions in the complex variable p (1–D case) over an arbitrary but fixed field of coefficients, is factorable as (matrix fraction description or representation)

$$T = AB^{-1} = D^{-1}C \tag{1.31}$$

where A, B, C, D are polynomial matrices over the ground field and A, B are right coprime while C, D are left coprime. The coprimeness conditions are equivalent to the requirement of existence of polynomial matrices X, Y, T, S with coefficients over the base field such that the respective conditions given below are satisfied. I denotes an identity matrix of appropriate order.

$$XA + YB = I \tag{1.32a}$$

$$CT + DS = I. \tag{1.32b}$$

It has been recently shown by Guiver and Bose [1.38] that matrices whose entries are rational functions in p_1, p_2 (2–D case) over a field of coefficients are factorable in a form similar to (3) with polynomial matrices A, B devoid of any nontrivial common right factor and polynomial matrices C, D devoid of any nontrivial common left factor. A, B, C, D are, of course, polynomial matrices in indeterminants p_1, p_2 (2–D case) over the arbitrary but fixed field. More importantly, the computations to extract a greatest common right divisor or a greatest common left divisor from two polynomial matrices having entries that are polynomials in p_1, p_2, need be performed in the specified ground field (and not in any extension field) containing the coefficients of the polynomial matrices (or the rational matrix from which the two initial non coprime polynomial matrices are derived). The primitive factorization algorithm which is central to the procedure under discussion cannot be entended to cases involving more than two indeterminates. Youla and Gnavi [1.39] delineated the various types of coprimeness (factor, minor, and zero) that are natural in multidimensional systems theory. These three types of coprimeness are equivalent in the 1–D case and are mutually distinct in the n–D case ($n > 2$). When $n = 2$, factor and minor coprimeness, are, interestingly, equivalent concepts and this fact justifies the feasibility of primitive factorization in the 2–D matrix case. Of course, the primitive factorization algorithm holds in the n–D case for any n, when attention is restricted to n-variate polynomials instead of polynomial matrices. Zero primeness imposes the most stringent restrictions on the $(m \times m)$ minors of a $m \times k, m \leq k$, polynomial matrix $A(p) \triangleq A(\mathbf{p}_1, p_2, \ldots, p_n)$. $A(\mathbf{p})$ is called zero-prime provided all its ($m \times m$) minors are devoid of common zeros, while it is minor-prime provided all its ($m \times m$) minors are devoid of common factors. Zero-primeness of $A(\mathbf{p})$ implies its unimodularity (and vice-versa); that is, zero primeness of $A(\mathbf{p})$ is equivalent to the existence of a polynomial matrix, $B(\mathbf{p})$ such that $A(\mathbf{p}) B(\mathbf{p})$ is the identity matrix of order m. Youla and Pickel [1.40], after defining that $A(\mathbf{p})$ is projectively free if it can be included at the first m rows of some ($r \times r$) elementary polynomial matrix showed how the Serre conjecture (proved independently by Quillen and Suslin as documented by Lam [1.43]) translates into the following fact. See also [1.89].

FACT 1.6. Any ($m \times r$) zero-prime polynomial matrix $A(\mathbf{p})$, $m \le r$ can be row-bordered up into a square ($r \times r$) elementary polynomial matrix $V(\mathbf{p})$. Moreover, if $A(\mathbf{p})$ is real, i.e. if all its entries are polynomials with real coefficients, $V(\mathbf{p})$ can also be constructed real.

Youla and Pickel proved Fact 1.6 using only a minimum of modern abstract algebra and justified the validity of the following equivalence.

Zero-prime ↔ Unimodular ↔ Projectively Free.

For polynomial matrices over an arbitrary commutative ring, validity of (1.32a) and (1.32b), imply, respectively, right and left zero (not, in general, factor or minor) coprimeness of polynomial matrices under consideration. In (1.32a), the polynomial matrices A, B are called right Bezout and in (1.32b) the polynomial matrices C and D are called left Bezout. Recently, Khargonekar [1.42], studied the existence and realization theory of matrix fraction representations in the forms given in (1.31) for linear systems over commutative rings. He was able to bring out the expected result that a transfer matrix T associated with an input/output map admits a left or right Bezout matrix fraction representation if and only if the map is split (see Fact 1.5). Split maps and zero-coprime fractional representations are, therefore, related.

Fuhrmann [1.43] developed a polynomial model approach to linear dynamical systems over a field with the objective of providing the exposition of state-space and matrix fraction representation schemes for realization, in a unified setting. He based his approach on representation theorems for submodules and quotient modules of spaces of polynomial matrices and vectors. A correspondence between fractional representations of the transfer matrix of a given input/output map and its reachable or observable realizations was established. Generalization of this approach to the case of systems over a principal ideal domain has been done by Conte and Perdon [1.44], whose results are documented in a more general setting in the work of Khargonekar, already referred to here.

Significant developments have taken place during the last decade in the consolidation, unification, and correlation of research on the realization theory over rings. These results directly apply to problems in the areas of multidimensional signal processing, delay-differential systems, realizations with parameters (Byrnes, [1.45]), realizations incorporating system robustness and parameter uncertainty, and, in general, in the domain of analysis and design of families of systems, instead of only particular systems. Often, realizations have to be obtained under severe constraints. This is, particularly, brought out in network theory, where the constraints of positive realness or restrictions on topology (like the doubly terminated lossless multidimensional two-port synthesis problem) can lead to serious, if not insurmountable problems. In fact, whether or not the multivariate positive realness property is sufficient for the synthesis of arbitrary multidimensional passive multiports remains an unresolved problem (see Bose, [1.5]). Multidimensional wave digital filters have a lot of attractive properties and it would be useful to provide general synthesis schemes for reference filters (see Fettweis, [1.15]. The stumbling blocks encountered in the realization of systems geared towards the types of applications referred provide impetus for further research and renewed challenge in an arena where new developments never seem to obliterate the scope and need for additional procedures, techniques, and refinements. Recent results which benefit the area of multivariate network realizability theory are summarized next.

1.3.2.1. *Integral Representation of Positive Real Functions*
Multivariate rational positive real functions and matrices form the nucleus in the study of multivariate network realizability theory.

DEFINITION 1.5. An $m \times m$ matrix $\mathbf{Z}(\mathbf{p})$ whose elements are (rational) functions in the complex variables $\mathbf{p} \triangleq (p_1, \ldots, p_n)$ is called (rational) positive real if
  (a) $\mathbf{Z}(\mathbf{p})$ is holomorphic in Re $\mathbf{p} > 0$
  (b) $[\mathbf{Z}^*(\mathbf{p})]' = \mathbf{Z}(p^*)$ in Re $\mathbf{p} > 0$
  (c) $\mathbf{Z}(\mathbf{p}) + [Z^*(\mathbf{p})]'$ is Hermitian nonnegative definite in Re $\mathbf{p} > 0$.
  Condition (a) implies that $\mathbf{Z}(\mathbf{p})$ need not be BIBO stable, condition (b) implies the realness of coefficients in the entries of the (rational) matrix

$Z(\mathbf{p})$, while condition (c) is a consequence of passivity of the network characterized by $Z(\mathbf{p})$. Positive real functions as well as their Poisson integral representations play a fundamental role in the theory of networks. In the $n = 1$ case, the following results are due to Cauer [1.56] who made use of the Riesz–Herglotz representation of a function, regular in $|z| < 1$, with its real part nonnegative in $|z| < 1$, and the bilinear transformation, $z = (p_1 - 1)/(p_1 + 1)$.

FACT 1.7. A function $Z(p_1)$ is positive real if and only if,

$$Z(p_1) = \int_{-\infty}^{\infty} \frac{jx\,p_1 - 1}{jx - p_1}\,d\mu(x) + cp_1,$$

where $\mu(x)$ is a real nondecreasing function of bounded variation on $(-\infty, \infty)$ and c is a nonnegative real constant. It is possible to select $\mu(x) = -\mu(-x)$, and the integral is taken to be in the Stieltjes sense.

FACT 1.8. [1.56]: Any positive real function, $Z(p_1)$ can be represented as,

$$Z(p_1) = p_1\left[c + \int_{0}^{\infty} \frac{d\mu(x)}{p_1^2 + x}\right].$$

where c is a nonnegative constant, μ is a nondecreasing function and the integral is to be taken in the Stieltjes sense. Conversely, if the preceding integral representation exists, it represents a positive real function.

For a representation theorem for positive real matrices in a single complex variable, which is the matrix counterpart of the result in Fact 1.7, see [1.57, Appendix 1]. More importantly, the multivariate counterpart of the Riesz–Herglotz result was given in 1963 [1.58].

FACT 1.9. [1.58]: The function $f: U^n \to C$ is holomorphic and has nonnegative real part in U^n if and only if it admits a representation,

$$f(\mathbf{z}) = j\,\mathrm{Im}\,f(\mathbf{0}) + \int_{T^n} \cdots \int \left[2 \prod_{k=1}^{n} \frac{1}{1 - z_k w_k^*} - 1\right] d\mu(\mathbf{w})$$

with a positive measure μ on T^n such that

$$\int_{T^n} \cdots \int w_1^{n_1} \ldots w_m^{n_m}\,d\mu(\mathbf{w}) = 0$$

unless

$$n_k \geq 0, \quad \text{for all} \quad k = 1, \ldots, m \qquad \text{or}$$

$$n_k \leq 0 \quad \text{for all} \quad k = 1, \ldots, m.$$

Note the orthogonality constraint that the positive measure μ must satisfy on T^n when $n > 1$. See also [1.90, pp. 80–91].

Aizenberg and Dautov [1.59] extended the above result for functions holomorphic in other polydomains besides U^n, but were not able to cover unbounded polydomains, which occur in the context of positive real functions or matrices. For these types of results relating to polydomains other than U^n, see [1.60]–[1.62], and [1.90] for more references.

1.3.2.2. *Sum of Squares Representation of a Form in Network Synthesis*

It has been seen (see Chapter 5 of [1.5]) that the infeasibility of sum of squares representation for a six variable positive (in the projective plane) form of degree 4 provides a theoretical limitation to a synthesis procedure for multivariate positive real matrices, which can characterize lumped-distributed and variable parameter multiports. Interestingly, this theoretical limitation also influences the extendability problem in multi-dimensional spectral estimation [1.46], and is linked to the fact that strongly positive [1.47] (see Fact 1.10 also) or completely positive [1.48] linear maps rather than positive linear maps are the natural generaliz-ation of positive linear functionals. Many examples exist to substantiate this fact and, quite recently, Schmüdgeon [1.49] gave the polynomial in (1.33) below as an example of a positive polynomial which is not a sum of squares of polynomials, and therefore it is a positive but not a strongly positive functional.

$$F(x_1, x_2) = 200(x_1^3 - 4x_1)^2 + 200(x_2^3 - 4x_2)^2 +$$
$$+ (x_2 - x_1)(x_2 + x_1)x_1(x_1 + 2)(x_1(x_1 - 2) + 2(x_2^2 - 4)). \tag{1.33}$$

In the context of network realizability theory, the class of functions synthesizable by Koga's procedure (see [1.5]) can, however, be identified from the theorem to be stated next.

Let $P_{n,m}$ be the set of all positive semidefinite forms in n variables of degree m. $P_{n,m}$ belongs to a closed convex cone in a finite dimensional Euclidean space. Let $P_{n,m}^r$ denote the family of members of $P_{n,m}$ in which all the intervening exponents are bounded by a given integer r i.e. if $F(x_1, x_2, \ldots, x_n) \in P_{n,m}^r$ then $F(x_1, x_2, \ldots, x_n)$ involves only monomials $x_1^{k_1} x_2^{k_2}$

... $x_n^{k_n}$ with $k_1 + k_2 + \ldots + k_n = m$ and $0 \le k_1, k_2, \ldots, k_n \le r$. Let $S_{n,m}$ be a subcone of $P_{n,m}$ formed from all finite sums of squares of polynomials. Clearly, $S_{n,m} \subseteq P_{n,m}$. Let $C[x_1, x_2, \ldots, x_n]$ be the algebra of all polynomials with complex coefficients in n commuting indeterminates $x_1, x_2, \ldots, x_n$. By a positive functional is meant a linear functional on $C[x_1, x_2, \ldots, x_n]$ which is nonnegative on $S_{n,m}$, $\forall n, m$. The functionals with nonnegative values on the cone $P_{n,m}$, $\forall n, m$, are strongly positive.

THEOREM 1.4. [1.50]: Suppose r, m are even and $m \ge 4$ ($m = 2$ case is trivial).
   (1) Let $n \ge 4$. Then, $P_{n,m}^r \cap P_{n,m} \subseteq S_{n,m}$,   iff $m \ge rn - 2$
   (2) Let $n = 3$. Then, $P_{n,m}^r \cap P_{n,m} \subseteq S_{n,m}$,
        iff $m = 4$,   or $m \ge 3r - 4$.
Since Koga's synthesis procedure depends upon the validity of $P_{n,m}^2 \cap P_{n,m} \subseteq S_{n,m}$ it only holds if and only if either $m \ge 2(n-1)$ or $m = 2$.

1.4. n–D PROBLEM OF MOMENTS AND ITS APPLICATIONS IN MULTIDIMENSIONAL SYSTEMS THEORY

This brief section is included to provide further scope for appreciation of the effects of distinction between the notions of positive and completely positive functionals in multidimensional problems. The problem of moments has an extensive literature. Shohat and Tamarkin [1.47] have given a concise but thorough documentation of the results in that area. Though the classical moment problem including its various ramifications like the trigonometric moment problem, and the Hausdorff moment problem is completely and satisfactorily solved in the 1–D case, the n–D ($n > 1$) counterpart provides some very interesting insights into the complexities encountered when attempting extensions of the 1–D results to several dimensions [1.51], [1.52]. The n–D moment problem is formulated as follows.

   PROBLEM FORMULATION: Let there be given an infinite multiple sequence of real constants

$$\mu_{i_1, i_2 \ldots i_n}, \qquad i_1, i_2, \ldots, i_n = 0, 1, 2, \ldots$$

in a n–D Euclidean space. The objective is to find necessary and sufficient conditions for a n–D distribution function $F(\mathbf{x})$ to exist (a distribution function is nonnegative, defined and finite over the family of all Borel sets in the n–D Euclidean space R^n and is completely additive) whose spectrum (defined as the set of all points $\mathbf{x}$ in R^n such that $F(I) > 0$ for every open set I containing $\mathbf{x}$) is to be contained in a closed set I_0, given in advance, and which is a solution to,

$$\mu_{i_1 \, i_2 \, \ldots \, i_n} = \int_{R^n} \left(\prod_{j=1}^{n} x_j^{ij} \right) dF(\mathbf{x}), \qquad (1.34)$$

for $i_1, i_2, \ldots, i_n = 0, 1, 2, \ldots$.

For the $n = 2$ case let $P(x_1, x_2)$ be any polynomial in x_1, x_2,

$$P(x_1, x_2) = \sum_{i_1, i_2} \alpha_{i_1} \beta_{i_2} x_1^{i_1} x_2^{i_2} \qquad (1.35)$$

where α_{i_1}, β_{i_2} are real or complex valued constants. Introduce the functional,

$$\mu(P) = \sum_{i_1, i_2} \mu_{i_1 \cdot i_2} \, \alpha_{i_1} \beta_{i_2} \qquad (1.36a)$$

where

$$\mu(x_1^{i_1} x_2^{i_2}) = \mu_{i_1 \cdot i_2}. \qquad (1.36b)$$

FACT 1.10. A necessary and sufficient condition that the I_0-moment problem defined by the sequence of moments $\{\mu_{i_1 i_2}\}$ shall have a solution is that the functional $\mu(P)$ be (I_0)-nonnegative i.e. $\mu(P) \geq 0$ whenever $P(x_1, x_2) \geq 0$ on I_0. When the functional $\mu(P)$ satisfies the preceding condition it is called strongly positive.

The preceding result suggests that for any positive polynomial $P(x_1, x_2)$ in (1.35), the linear functional T defined by

$$\left\langle T, \sum_{i_1=0} \sum_{i_2=0} \alpha_{i_1} \beta_{i_2} x_1^{i_1} x_2^{i_2} \right\rangle \triangleq \sum_{i_1=0} \sum_{i_2=0} \mu_{i_1 i_2} \, \alpha_{i_1} \beta_{i_2}$$

$$= \sum_{i_1=0} \sum_{i_2=0} \alpha_{i_1} \beta_{i_2} \iint x_1^{i_1} x_2^{i_2} \, dF(x_1, x_2)$$

$$= \iint P(x_1, x_2) \, dF(x_1, x_2)$$

is positive. T is multiplicatively positive if $\langle T, \phi \phi^* \rangle \geq 0$ for every test function $\phi(x_1, x_2)$. In the linear space of all real polynomials in variables x_1, x_2, a multiplicatively positive functional may not be positive and therefore the condition $\langle T, \phi^2(x_1, x_2)\rangle \geq 0$ for all real polynomials $\phi(x_1, x_2)$ is not sufficient for the solvability of the 2–D moment problem.

In the trigonometric moment problem, I_0 is the distinguished boundary T^n of a polydisc and (1.34) is replaced by,

$$\mu_{i_1 i_2 \ldots in} = \mu(\mathbf{i}) = \int \cdots \int_{T^n} e^{j\langle \mathbf{i}, \mathbf{z}\rangle} \, dF(\mathbf{z}) \qquad (1.37)$$

where $\mathbf{z} = (z_1, \ldots, z_n), \mathbf{i} = (i_1, \ldots, i_n)$ and $\langle \mathbf{i}, \mathbf{z}\rangle = i_1 z_1 + \ldots + i_n z_n$.

A representation similar to the one given above occurs in probability theory [1.53] and the multidimensional spectrum estimation problem [1.46]. It is also pointed out that the notion of strong positivity has important consequences in the theory of unbounded operators [1.54]. In the problem of multidimensional spectrum estimation occurring in diverse fields of applications of the theory of multidimensional signal processing (1.37) relates a specified finite set of correlation samples $\{\mu(\mathbf{i}): \mathbf{i} \in \Delta = (0, \pm \delta_1, \ldots, \pm\delta_m), \delta_i \in Z, i = 1, 2, \ldots, m)\}$ to a real positive spectrum $F(\mathbf{x}), \mathbf{x} \in D^n \subset R^n$, to be estimated, provided it exists, such that for $\mathbf{i} \in \Delta$,

$$\mu(\mathbf{i}) = \int \cdots \int_{D^n} F(\mathbf{x}) \, e^{j\langle \mathbf{i}, \mathbf{x}\rangle} \, d\mathbf{x}, \qquad (1.38)$$

where D^n is the domain over which $F(\mathbf{x})$ is specified to be nonzero. If a solution to (1.38) for a positive $F(\mathbf{x})$ exists, then the specified correlation samples are said to be extendible. Without specializing $\mathbf{i}$ to take values only on the finite Δ, $\mu(\mathbf{i})$ in (1.38) may be viewed as a continuous function of $\mathbf{i}$ which is positive definite in the following sense. For any points $\mathbf{i}_1, \mathbf{i}_2, \ldots \mathbf{i}_m$ in R^n and any complex numbers $\xi_1, \ldots, \xi_m$ one has

$$\sum_{k=1}^{m} \sum_{j=1}^{m} \mu(\mathbf{i}_k - \mathbf{i}_j) \, \xi_k \xi_j^* \geq 0 \qquad (1.39a)$$

for every positive integer m, where

$$\mathbf{i}_k - \mathbf{i}_j \overset{\Delta}{=} (i_{k1} - i_{j1}, \ldots, i_{kn} - j_{kn}). \qquad (1.39b)$$

Indeed (1.39a) follows from the use of (1.38) in the following manner.

$$\sum_{k=1}^{m} \sum_{r=1}^{m} \mu(\mathbf{i}_k - \mathbf{i}_r) \, \xi_k \xi_r^* = \sum_{k=1}^{m} \sum_{r=1}^{m} \xi_k \xi_r^* \times$$

$$\times \int_{D^n} \cdots \int_{-\infty}^{\infty} e^{j\langle i_k - i_r, \, \mathbf{x}\rangle} \, F(\mathbf{x}) \, d\mathbf{x} = \int_{D^n} \cdots \int_{-\infty}^{\infty} \Big| \sum_{r=1}^{m}$$

$e^{j\langle i_r, \, \mathbf{x}\rangle} \, \xi_r|^2 \, F(\mathbf{x}) \, d\mathbf{x} \geq 0$, since $F(\mathbf{x})$ is positive.

In the 1–D case, for a specified finite set of samples $\mu(\mathbf{i})$, $\mathbf{i}\epsilon\Delta = \{0, \pm \delta_1, \ldots, \pm\delta_M\}$, the positive definitess of the Toeplitz matrix with $[\mu(0) \, \mu(\delta_1) \ldots \mu(\delta_M)]^t$ in the first row and $[\mu(0) \, \mu(-\delta_1) \ldots \mu(-\delta_M)]^t$ in the first column is necessary and sufficient for the extendability of $(\{\mu(\mathbf{i})\}: \mathbf{i}\epsilon\Delta)$. This can be substantiated via results of Rudin. Rudin [1.55] considered a finite set S in a discrete group G and related the extension problem to the sum of squares representation problem of certain positive trigonometric polynomials. He then showed that the problem of extending the class of all continuous complex-valued functions μ on $S - S$ (the set of all points $i_k - i_r \in G$, with $i_k\epsilon S$ and $i_r\epsilon S$) which satisfies the 1–D (finite S) counterpart of (1.39a) has a solution if $G = Z$, the additive group of integers and if S is a finite arithmetic progression in G. Rudin also proved that the analogous result fails to hold in higher dimensions i.e. when $n > 1$. In fact, he showed that not all positive trigonometric polynomials on T^2 are representable as sums of squares of trigonometric polynomials and used this fact to demonstrate that the extension problem may fail to have a solution if $G = Z \times Z \overset{\Delta}{=} Z^2$, the group of all lattice points in the plane and if S is a square of lattice points. Subsequently, Rudin transferred the result on infeasibility, in general, of extension from Z^2 to R^2 and hence to R^n, $n \geq 2$. A Toeplitz form, like in (1.39a), on $S \times S$, where S is a subset of, say $G = Z^n = Z \times Z \times \ldots \times Z$ is a function, μ on the Cartesian product $S \times S$ with the property that $\mu(i_k, i_j)$ is a function only of $i_k - i_j$, $i_k\epsilon S$, $i_j \epsilon S$. Then, the preceding discussion centering around the validity of (1.39a) as a consequence of the positive definiteness property of forms induced by positive measures via the representation given in (1.37) or (1.38), together with Rudin's results, lead to the following important fact.

FACT 1.11. [1.67]: Every Toeplitz form induced by a positive measure is positive definite but the converse is false if $n > 1$.

Fact 1.11 imposes serious limitations on the multidimensional spectral estimation problem, where a procedure for verifying whether or not a finite multidimensional Toeplitz form is the restriction of a form induced by a positive measure is required. One procedure to attain this objective has been reported in [1.67, pp. 362–363].

1.5. ROLE OF IRREDUCIBLE POLYNOMIALS IN
MULTIDIMENSIONAL SYSTEMS THEORY

The role of irreducible polynomials has been noted, from the very beginning, in [1.5] and these types of polynomials play a crucial role in the
problem of unambiguous determination of a function (object) from the
modulus of its Fourier transform (energy spectra) or equivalently from its
autocorrelation function. This problem is important at high frequency
applications like optics where the only experimentally measurable
quantity is the scattered-intensity distribution. In the discrete case, i.e.
when the brightness distribution is specified in a rectangular array, the
image reconstruction problem reduces to the determination of a poly-
nomial (uniquely, if possible) $p(\mathbf{z}) \triangleq p(z_1, z_2, \ldots z_n)$ from a specified
knowledge of the product (in (1.40) $\mathbf{z}^{-1} \triangleq (z_1^{-1}, z_2^{-1}, \ldots, z_n^{-1})$)

$$A(\mathbf{z}, \mathbf{z}^{-1}) = p(z_1, z_2, \ldots, z_n) \, p(z_1^{-1}, z_2^{-1}, \ldots, z_n^{-1}) \qquad (1.40)$$

subject to the physical constraint that the *coefficients* (representing the
brightness distribution) of $p(z_1, z_2, \ldots, z_n)$ *are nonnegative*. The question
of uniqueness of $p(z_1, \ldots, z_n)$ is related to its irreducibility. See 'Open
Problems', Problem #2 for more on uniqueness of reconstruction of a
multidimensional sequence from either the phase or magnitude of its
Fourier transform.

The coefficients of $A(\mathbf{z})$ form the sequence of discrete autocorrelation
function of the unknown image required to be restored. In the 1–D case,
the brightness distribution cannot, in general, be reconstructed un-
ambiguously. Uniqueness of reconstruction up to trivial factors (like
multiplication by a positive real constant, and turning of the brightness
distribution by 180°) is possible when the polynomial $p(\mathbf{z})$ turns out to be
irreducible [1.68]. Carlitz [1.69] gave an asymptotic formula for the
number of irreducible polynomials in several variables of total degree m
and arrived at the conclusion that almost all multivariate polynomials are
irreducible (for this, see also [1.70]). For enumeration of irreducible
multivariate polynomials whose coefficients are in Z_q, q prime, see [1.5].
These results justify the reduced ambiguity of the image reconstruction
problem in the discrete multidimensional case. Manolitsakis [1.71] considered the reconstruction problem of two-dimensional continuous
objects from their energy spectra. He showed that the properties of
two-dimensional fields differ inherently from those of one-dimensional
fields (essentially because the zero-sets are not isolated) and that they

lead to a reduced ambiguity for object reconstruction from intensity data. See 'Open Problems', Problem #3 for more research into conditions on polynomial coefficients required to guarantee irreducibility as required in some applications.

Irreducible polynomials with coefficients over finite field Z_q, where q is a prime integer are known to play useful roles in the construction of arrays with special properties, which are generalizations of some of the properties of pseudorandom sequences (see Chapter 6 of [1.5]). In the 1–D case particular types of irreducible polynomials called primitive polynomials are used to design linear feedback shift registers for generating pseudorandom sequences, which have the two-level discrete auto-correlation property when $q = 2$. Binary arrays with a maximum of three distinct levels of autocorrelation have been constructed in [1.5] and explicit expressions for the values of these levels have been given. In Z_q, q prime, arrays with not more than q distinct levels of autocorrelation are discussed in [1.72], where however, a distinction between the notions of period and the maximum area property of linear arrays from their counterparts in the case of pseudorandom sequences are drawn. A natural generalization of pseudorandom (maximum period) sequences to maximum period arrays is given in [1.73], where, however, the concept of period is different from that in [1.5] and [1.72]. The delineation of properties of arrays studied in [1.73] have been motivated by applications in 2–D cyclic code construction and such constructive schemes have been subsequently considered in [1.74].

1.6. HILBERT TRANSFORM AND SPECTRAL FACTORIZATION

The problem of determining the phase of a certain function analytic in one half of the complex plane from the knowledge of its magnitude on the axis bounding the half planes has been of interest in circuit theory, communication theory (in the context of modulation and demodulation of signals) and in various branches of physics including the theory of decay of elementary particles, X-ray crystallography, coherence theory, and light scattering. The theory of analytic functions via the Hilbert transform pair (dispersion relations) technique provides a relationship between the real and imaginary parts or the magnitude and phase of certain analytic functions (for the magnitude-phase relationships to hold the analytic function should also have no zeros in the half-plane of interest i.e. the function is minimum phase). It has been pointed out in

[1.5] that problems occur in trying to extend the 1–D Hilbert transform relationships to the multidimensional setting. In [1.75], this fact is substantiated by demonstrating that the 2–D dispersion relations constitute a pair of inhomogenous singular integral equations with no solution.

Another limitation in the applications of multidimensional methods occur in the inability to factor a positive (on the multi-axis or poly-boundary) function in the form shown in (1.40) subject to the constraint that the zero-set of the finite order polynomial factor be constrained in a half-plane. In the discrete case, the zero-set of a spectral factor should be excluded from a unit polydisc. Note that zero sets of multivariate polynomials form continuous algebraic curves which extend throughout the space, i.e. cannot be constrained to belong to bounded polydomains. Helson and Lowdenslager [1.76], however, defined half-plane spectral factors, which satisfy stability constraints, but, in general have infinite support. Due to greater versatility of a half-plane filter frequency response over that of a quarter-plane filter (fortunately the theoretical foundations of half-plane filter design are well established in contrast to the quarter-plane case), half-plane filters are being increasingly used in multidimensional filtering, prediction [1.77] and spectrum estimation problems [1.78], [1.46].

1.7. CONCLUSIONS

The progress and directions of research in several areas of multidimensional systems theory since about 1981 has been reported. In the process, some open problems have been brought to the attention of the reader. The breadth of applications of the theoretical results discussed is, indeed, impressive, and we list below some additional specific areas of applications that have not, yet, been cited in this chapter. First, in addition to the flurry of activities associated with spatio-temporal signal processing [1.91], optical and electronic feedback systems (these systems incorporating spatial and temporal variables are called hybrid systems) have been proposed for various purposes like iterative image processing and image restoration. For the state-space representation and input/output description of spatio-temporal systems, analysis of internal stability, and, discussion of the related topic of stabilizability, read Chapter 4 by Kamen. Chapter 3 by Guiver and Bose considers the problem of stabilization of 2–D systems by causal and weakly causal compensators and contains references to the recent literature on spatio-temporal feedback systems.

Second, a variety of applications, especially in image restoration problems, has motivated research into multidimensional linear shift-variant systems. In Chapter 5, Valenzuela and Bose document known as well as new results in this area. Stability conditions for 1–D causal linear shift-variant systems have been shown to be similar in content to the well established theorems on the BIBO stability of 2–D LSI systems [1.79]. Third, stability results for n-dimensional linear shift invariant digital filters also become applicable in stability investigations of externally bilinear systems (which form a subclass of nonlinear systems) whose output response is a bilinear function of two inputs, each applied to a different input terminal. Kamen has shown that a n–D shift invariant externally bilinear system is BIBO stable if a particular member of the associated class of linear shift invariant (LSI) $2n$–D system is BIBO stable, but not vice-versa [1.80].

In Section 1.3, the thrust towards research into the realization theory of linear systems over commutative rings was evident. This was to a large extent motivated by applications of the ring structure in systems characterizable by delay-differential equations when those are interpretable as differential equations whose coefficients belong to a polynomial ring in one (in case of commensurate delays) or several variables (in case of incommensurate delays) [1.81], in image processing or 2–D digital filtering systems (see Chapter 3), and in the global study of families of linear systems initiated in [1.82]. Besides questions related to realization, other questions can also be meaningfully posed about linear systems over rings. The answers to these questions, however, are usually difficult to obtain and often the use of new tools and methods of approach are required in comparison with those used to tackle analogous problems for classical linear systems over fields. For results, conjectures, and open problems on 'pole-shifting' over rings, see [1.83]. An excellent survey of various recent results on linear systems over commutative rings was prepared by Sontag [1.84].

Multidimensional linear passive systems are frequently encountered in mathematical physics and electrical engineering. The nucleus to the study of such systems is provided by the concept of positive realness. Multivariate positive real matrices occurring in network theory characterize only a subclass of passive systems. In a more general context passivity of an operator can be defined relative to an acute, closed, convex solid cone in R^n with vertex at zero. Then, a large number of physical processes describable by partial differential equations are passive systems relative to certain cones. A general theory of shift-invariant multidimensional systems based on the theory of multivariate positive real matrices has been presented by Vladimorov [1.63]. Aside from electrical network theory, 1–D linear passive system theory is known to be useful in the descriptions of thermodynamic systems and the scattering theory of electromagnetic waves and elementary particles. Vladimorov gave an exposition of the considerably increasing scopes for applications of multidimensional linear passive systems theory and noted the relevance of the fundamental concept of positive realness (with respect to an appropriate cone) in equations of the theory of elasticity, magnetohydrodynamics, the equations of rotating fluid and acoustics, Dirac's equation in quantum physics and the transfer equation. Subsequently, Drožžinov [1.64] investigated a class of first order partial differential equations with infinitely differentiable coefficients and identified the subclass of systems that become passive relative to a cone. In [1.65], Drožžinov also proved some results on the quasi-asymptotic behaviour of fundamental solutions of passive shift-invariant systems after developing a multidimensional Tauberian theorem for holomorphic functions of bounded arguments. Note that the argument of any positive real function is continuous and bounded in the polydomain of holomorphy.

In Section 1.4, the multidimensional moment problem has been considered. Besides the applications already cited, the n–D moment problem, when $n = 2$, is useful in image reconstruction [1.85], where given some finite set of moments,

$$M(j, k) = \iint f(x, y) \, x^j y^k \, dx \, dy$$

of an image plane irradiance distribution, $f(x, y)$, the object is to determine how well the image can be reconstructed. In Sections 1.5 and 1.6, respectively, the roles of irreducible polynomials in multidimensional systems theory and the consequence of problems in extending 1–D Hilbert transform and spectral factorization results to the multidimensional setting are briefly summarized.

There are other mathematical limitations in the extension of 1–D results to the n–D ($n > 1$) case which influence applications of the theory of multidimensional systems. For example, it is known that in the unit disc in the complex plane, every function f in the Nevanlinna class N has a factorization, $f = I \times E$, where I is inner and E is outer. However, there exist bounded analytic functions on a polydisc (or for that matter on a ball), that have no inner-outer factorization [1.6]. Rubel [1.86] introduced the notions of internal and external analytic functions of several complex variables and proved that on any simply-connected complex analytic manifold of any finite dimension every f in the RP-Nevanlinna class, $RP{:}N$, (the class $RP{:}N$ is defined as the usual Nevanlinna class N, except that one uses pluriharmonic majorants rather than n-harmonic majorants), has a factorization, $f = I \times E$, where I is internal and E is external. The notion of weakly inner-strongly outer factorization has been used in [1.87] to study the asymptotic behaviour of planar least-squares inverse polynomials occurring in studies of stabilization of 2–D digital filters.

In order to satisfy constraints of size this chapter, and for that matter, this book is concerned with tools relevant to the study of deterministic multidimensional systems. Random fields or distributed disordered systems, characterized by random variation over space and time are not considered and the reader is referred to [1.88] for information on two-dimensional and multidimensional local average processes.

REFERENCES

[1.1] N. K. Bose (ed.), Special Issue on *Multidimensional Systems*, Proceedings of IEEE, Vol. 65, June 1977.

[1.2] M. G. Ekstrom and S. K. Mitra (eds.), *Two-Dimensional Signal Processing*, Dowden, Hutchinson and Ross, New York, 1978.

[1.3] N. K. Bose (ed.), *Multidimensional Systems: Theory and Applications*, IEEE Press, New York, 1979.

[1.4] A. S. Willsky, *Digital Signal Processing and Control and Estimation Theory: Points of Tangency, Areas of Intersection, and Parallel Directions*, The MIT Press, Cambridge, Massachusetts, 1979.

[1.5] N. K. Bose, *Applied Multidimensional Systems Theory*, Van Nostrand Reinhold, New York, 1982.

[1.6] W. Rudin, *Function Theory in Polydiscs*, W. A. Benjamin Inc., New York, 1969.

[1.7] Sh. A. Dautov, 'Some Questions of Multidimensional Complex Analysis', Akad. Nauk SSSR Sibirisk, Otdel. Inst. Fiz. Krasnoyarsk, 1980, p. 19 (in Russian; translation help provided to the author by Professor Y. Tsypkin through the courtesy of Professor E. I. Jury).

[1.8] Sh. A. Dautov, 'On Absolute Convergence of the Series of Taylor Coefficients of a Rational Function of Two Variables: Stability of Two-Dimensional Recursive Digital Filters', *Soviet Math. Dokl.*, 23, No. 2, 1981 (American Mathematical Society Translations), pp. 448–451.

[1.9] D. Goodman, 'Some Stability Properties of Two-dimensional Linear Shift-invariant Digital Filters', *IEEE Trans. Circuits and Systems*, 24, 1977, pp. 201–208.

[1.10] R. K. Alexander and J. W. Woods, '2–D Digital Filter Stability in the Presence of Second-kind Nonessential Singularities', *IEEE Trans. CAS*, 29, Sept. 1982, pp. 604–612.

[1.11] I. E. Zak, 'On Absolute Convergence of Double Fourier Series', *Soobsceniya Akad. Nauk Gruzin. SSR*, 12, 1951, pp. 129–133.

[1.12] P. Delsarte, Y. V. Genin, and Y. V. Kamp, 'A Simple Proof of Rudin's Multivariable Stability Theorem', *IEEE Trans. Acoustics, Speech and Signal Proc.*, 28, Dec. 1980, pp. 701–705.

[1.13] E. I. Jury, 'Stability of Multidimensional Systems and Related Problems', Chapter 9 of forthcoming book on *Progress in Multidimensional System Theory*, edited by S. G. Tzafestas, Marcel Dekker, Boston, MA.

[1.14] T. S. Huang (ed.), *Two-Dimensional Digital Signal Processing 1 Linear Filters*, Vol. 42, Topics in Applied Physics, Springer-Verlag, New York, 1981.

[1.15] A. Fettweis, 'On the Scattering Matrix and the Scattering Transfer Matrix of Multidimensional Lossless Two-parts', *Archiv für Elektronik und Übertragungstechnik*, 36, 1982, pp. 374–381.

[1.16] I. Gregor, 'Biquadratic *n*-Dimensional Impedances', *Int. J. Circuit Theory and Applications*, 9, 1981, pp. 369–377.

[1.17] P. Delsarte, Y. V. Genin, and Y. V. Kamp, 'An Equivalence between Bounded Multivariable Functions and a Class of Bounded Single Variable Functions', *Int. J. Control*, 34. 1981, pp. 383–389.

[1.18] P. Delsarte, Y. V. Genin, and Y. V. Kamp, 'Koga's Multivariable Stability Criterion', *Proc. of IEEE*, 70, March 1982, pp. 298–299.

[1.19] W. E. Brumley, 'On the Asymptotic Behavior of Solutions of Differential-Difference Equations of Neutral Type', *J. Diff. Equations*, 7, 1970, pp. 175–188.

[1.20] P. S. Gromova, 'Stability of Solutions of Nonlinear Equations of the Neutral Type in the Asymptotically Critical Case', *Mathematical Notes of the Academy of Sciences of the USSR*, Vol. 1, No. 5, May–June 1967, pp. 472–479. (Translated from Mathematicheskie Zametki, Vol. 1, No. 6, June 1967, pp. 715–726).

[1.21] J. P. Guiver and N. K. Bose, 'On Test for Zero-sets of Multivariate Polynomials in Noncompact Polydomains', *Proc. of IEEE*, 69, April 1981, pp. 467–469.

[1.22] Y. V. Genin, 'An Algebraic Approach to *A*-stable Linear Multistep Multiderivative Integration Formulas', *BIT*, 14, 1974, pp. 382–406.

[1.23] P. Delsarte, Y. V. Genin, and Y. V. Kamp, 'A Proof of the Daniel–Moore Conjectures for *A*-stable Multistep Two-Derivative Formulae', *Phillips J. Res.*, 36, 1981, pp. 79–88.

[1.24] G. Dahlquist, 'Positive Functions and Some Applications to Stability Questions for Numerical Methods', in de Boor, C. and Golub, G. (eds.), *Recent Advances in Numerical Analysis*, Academic Press, New York, 1978, pp. 1–29.

[1.25] J. B. Edwards and D. H. Owens, *Analysis and Control of Multipass Processes*, Research Studies Press, a division of John Wiley and Sons, Ltd., Chichester, England, 1982.

[1.26] E. W. Kamen, 'Linear Systems with Commensurate Time Delays: Stability and Stabilization Independent of Delay', *IEEE Trans. Automatic Control*, Vol. 27, April 1982, pp. 367–375; also corrections in *IEEE Trans. Automatic Control*, Vol. 28, Feb. 1983, pp. 248–249.

[1.27] Y. Rouchaleau, B. E. Wyman, and R. E. Kalman, 'Algebraic Structure of Linear Dynamical Systems: III Realization Theory over a Commutative Ring', *Proc. Nat. Acad. Sci.*, U.S.A., 69, 1972, pp. 3404–3406.

[1.28] Y. Rouchaleau and E. D. Sontag, 'On the Existence of Minimal Realizations of Linear Systems over Noetherian Integral Domains', *J. Computer and Systems Science*, 18, Feb. 1979, pp. 65–75.

[1.29] E. D. Sontag, 'The Lattice of Minimal Realizations of Response Maps over Rings', *Math. Systems Theory*, 11, 1977, pp. 169–175.

[1.30] E. D. Sontag, 'On Split Realizations of Response Maps over Rings', *Information and Control*, 37, April 1978, pp. 23–33.

[1.31] E. D. Sontag, 'Linear Systems over Commutative Rings: A Survey', *Ricerche di Automatica*, 7, 1976, pp. 1–34.

[1.32] R. Eising, 'State-space Realization and Inversion of 2–D Systems', *IEEE Trans. Circuits and Systems*, 27, July 1980, pp. 612–619.

[1.33] S. G. Tzafestas and T. G. Pimenides, 'Feedback Characteristic Polynomial Controller Design of 3–D Systems in State-Space', *J. of the Franklin Institute*, 314, Sept. 1982, pp. 169–189.

[1.34] S. G. Tzafestas and T. G. Pimenides, 'Feedback Decoupling-controller Design of 3–D Systems in State-space', *Mathematics and Compuiers in Simulation*, 24, 1982, pp. 341–352.

[1.35] M. Fliess, 'Sur la réalisation des systémes dynamiques bilineaires', *C. R. Acad. Sci. Paris* 227A, 1973, pp. 923–926.

[1.36] E. D. Sontag, 'On Linear Systems and Noncommutative Rings', *Math. Systems Theory*, 9, 1976, pp. 327–344.

[1.37] E. Fornasini, 'On the Relevance of Noncommutative Power Series in Spatial Filter Realization', *IEEE Trans. Circuits and Systems*, 25, May 1978, pp. 290–299.

[1.38] J. P. Guiver and N. K. Bose, 'Polynomial Matrix Primitive Factorization over Arbitrary Coefficient Field and Related Results, *IEEE Trans. Circuits and Systems*, 10, Oct. 1982, pp. 649–657.

[1.39] D. C. Youla and G. Gnavi, 'Notes on n-Dimensional System Theory', *IEEE Trans. Circuits and Systems*, 26, Feb. 1979, pp. 105–111.

[1.40] D. C. Youla and P. F. Pickel, 'The Quillen-Suslin Theorem and the Structure of n-Dimensional Elementary Polynomial Matrices', Poly–MRI–1422–82, Polytechnic Institute of New York, Farmingdale, New York, 1982. Also in *IEEE Trans. Circuits and Systems*, 31, June 1984, pp. 513–518.

[1.41] T. Y. Lam, *Serre's Conjecture*, Lecture Notes in Mathematics 635, Springer-Verlag, New York, 1978.

[1.42] P. P. Khargonekar, 'On Matrix Fraction Representations for Linear Systems over Commutative Rings', *SIAM J. Control and Optimization*, 20, March 1982, pp. 172–197.

[1.43] P. A. Fuhrmann, 'Algebraic Systems Theory: An Analyst's Point of View', J. Franklin Institute, 305, 1976, pp. 521–540.

[1.44] G. Conte and P. M. Perdon, 'Systems over a Principle Ideal Domain: A Polynomial Model Approach', *SIAM J. Control and Optimization*, 20, Jan. 1982, pp. 112–124.

[1.45] C. I. Byrnes, 'On the Control of Certain Deterministic Infinite-dimensional Systems by Algebro-geometric Techniques', *Amer. J. Math.*, 100, 1978, pp. 1333–1381.

[1.46] J. H. McClellan, 'Multidimensional Spectral Estimation', *Proc. of IEEE*, 70, Sept. 1982, pp. 1029–1039.

[1.47] J. A. Shohat and J. D. Tamarkin, *The Problem of Moments*, Amer. Math. Soc., Mathematical Surveys, No. 1, N.Y., 1943.

[1.48] W. F. Stinespring, 'Positive Functions on C^*-algebras', *Proc. Amer. Math Soc.*, 6, 1955, pp. 211–216.

[1.49] K. Schmüdgeon, 'An Example of a Positive Polynomial which is not a Sum of Squares of Polynomials – a Positive but not Strongly Positive Functional', *Math. Nachr.*, 88, 1979, pp. 385–390.

[1.50] M. D. Choi and T. Y. Lam, 'External Positive Semidefinite Forms', *Math. Ann.*, 231, 1977, pp. 1–18.

[1.51] E. K. Haviland, 'On the Momentum Problem for Distributions in more than One Dimension 1', *Amer. J. Math.*, 57, 1935, pp. 562–568.

[1.52] E. K. Haviland, 'On the Momentum Problem for Distribution Functions in more than One Dimension 2', *Amer. J. Math.*, 58, 1936, pp. 164–168.

[1.53] H. Cramer, *Random Variables and Probability Distributions*, Cambridge Tract in Mathematics and Mathematical Physics, No. 36, London, 1937.

[1.54] R. T. Powers, 'Self-adjoint Algebras of Unbounded Operators', *Comm. Math. Phys.*, 21, 1971, pp. 85–124.

[1.55] W. Rudin, 'The Extension Problem of Positive Definite Functions', *Illinois J. Maths*, 7, 1963, pp. 532–539.

[1.56] W. Cauer, 'The Poisson Integral for Functions with Positive Real Part', *Bull. Amer. Math. Soc.*, 38, 1932, pp. 713–717.

[1.57] E. J. Beltrami and M. R. Wohlers, *Distributions and the Boundary Values of Analytic Functions*, Academic Press, New York, 1966.

[1.58] A. Korányi and L. Pukánszky, 'Holomorphic Functions with Positive Real Part on Polycylinders', *Trans. Amer. Math. Soc.*, 108, 1963, pp. 449–456.

[1.59] L. A. Aizenberg and Š. A. Dautov, 'Holomorphic Functions of Several Variables with Nonnegative Real Parts; Traces of Holomorphic and Pluriharmonic Functions on the Šilov Boundary', Mat. Sb. U.S.S.R., 28, No. 3, pp. 301–313.

[1.60] A. Korányi, 'The Poisson Integral for Generalized Half-planes and Bounded Symmetric Domains', *Annals of Mathematics*, Vol. 82 (2nd series), 1965, pp. 332–350.

[1.61] V. S. Vladimorov, 'Holomorphic Functions with Nonnegative Imaginary Part in a Tubular Domain over a Cone', *Math. U.S.S.R. Sbornik*, 8, No. 1, 1969, pp. 126–146.

[1.62] V. S. Vladimorov, 'Holomorphic Functions with Positive Imaginary Part in the Future Tube I and II', *Math. U.S.S.R. Sbornik*, 22, 1974, pp. 1–16 (*Math. U.S.S.R. Sbornik*, 23, No. 4, 1976, pp. 467–482).

[1.63] V. S. Vladimorov, *Generalized Functions in Mathematical Physics*, 2nd ed., 'Nauka', Moscow, 1979; English transl., 'Mir', Moscow, 1979.

[1.64] J. U. N. Drožžinov, 'Passive Linear Systems of Partial Differential Equations', *Math. U.S.S.R. Sbornik*, 44, No. 3, 1983, pp. 269–278.

[1.65] J. U. N. Drožžinov, 'A Multidimensional Tauberian Theorem for Holomorphic Functions of Bounded Argument and the Quasi-asymptotics of Passive Systems', *Math. U.S.S.R. Sbornik*, 45, No. 1, 1983, pp. 45–61.

[1.66] S. M. Nikolskii, *Approximation of Functions of Several Variables and Imbedding Theorems*, Springer-Verlag, Berlin, 1975.

[1.67] G. Cybenko, 'Moment Problems and Low Rank Toeplitz Approximations', *Circuits, Systems, and Signal Processing*, Vol. 1, Nos. 3–4, 1982, pp. 345–366.

[1.68] Y. M. Bruck and L. G. Sodin, 'On the Ambiguity of the Image Reconstruction Problem', *Optical Communications*, 30, Sept. 1979, pp. 304–308.

[1.69] L. Carlitz, 'The Distribution of Irreducible Polynomials in Several Indeterminates', *Ill. J. Math.*, 1963, pp. 371–375.

[1.70] M. M. Hayes and J. H. McClellan, 'Reducible Polynomials in more than One Variable', *Proc. IEEE*, 70, Feb. 1982, pp. 197–198.

[1.71] I. Manolitsakis, 'Two-dimensional Scattered Fields: A Description in Terms of the Zeros of Entire Functions', *J. Math. Phys.*, 23, Dec. 1982, pp. 2291–2298.

[1.72] T. Nomura, H. Miyakawa, H. Imai, and A. Fukuda, 'A Theory of Two-dimensional Linear Recurring Arrays', *IEEE Trans. Inform. Theory*, 18, Nov. 1972, pp. 775–785.

[1.73] S. Sakata, 'General Theory of Doubly Periodic Arrays over an Arbitrary Finite Field and Its Applications', *IEEE Trans. Inform. Theory*, 24, Nov. 1978, pp. 719–730.

[1.74] S. Sakata, 'On Determining the Independent Point Set for Doubly Periodic Arrays and Encoding Two-dimensional Cyclic Codes and Their Duals', *IEEE Trans. Inform. Theory*, 27, Sept. 1981, pp. 556–565.

[1.75] M. Nieto Vesperinas, 'Dispersion Relations in Two Dimensions: Application to the Phase Problem', *Optik*, 56, 1980, pp. 377–384.

[1.76] H. Helson and D. Lowdenslager, 'Prediction Theory and Fourier Series in Several Variables', *Aita Math*, 99, 1958, pp. 165–202.

[1.77] T. L. Marzetta, 'Two-dimensional Linear Prediction: Autocorrelation Arrays, Minimum-phase Prediction Error Filters and Reflection Coefficient Arrays', *IEEE Trans. Acoustics, Speech and Signal Processing*, 28, Dec. 1980, pp. 725–733.

[1.78] T. L. Marzetta, 'Additive and Multiplicative Minimum-phase Decompositions of 2–D Rational Power Density Spectra', *IEEE Trans. Circuits and Systems*, 29, April 1982, pp. 207–214.

[1.79] D. J. Schmidlin, 'The Characterization of Causal Shift-variant Systems Excited by Causal Inputs', *IEEE Trans. Circuits and Systems*, 28, Oct. 1981, PP. 981–994.

[1.80] E. W. Kamen, 'On the Relationship between Bilinear Maps and Linear Two-dimensional Maps', *Nonlinear Analysis: Theory, Methods and Applications*, 3, 1979, pp. 467–481.

[1.81] N. S. Williams and V. Zakian, 'A Ring of Delay Operators with Applications to Delay Differential Systems', *SIAM J. Control and Optimization*, 15, 1977, pp. 247–255.

[1.82] M. Hazewinkel and R. E. Kalman, 'Invariants, Canonical Forms and Modulii for Linear Constant Finite Dimensional Dynamical Systems', *Lecture Notes in Economics and Math. Systems*, 131, 1976, pp. 48–60.

[1.83] R. Bumby, E. D. Sontag, H. J. Sussman, and W. Vasconcelos, 'Remarks on the Pole-shifting Problem over Rings', *J. Pure and Applied Algebra*, 20, 1981, pp. 113–127.

[1.84] E. D. Sontag, 'Linear Systems over Commutative Rings: A (Partial) Updated Survey', *Proc. IFAC, Kyoto, Japan*, Aug. 1981.

[1.85] M. R. Teague, 'Image Analysis via the General Theory of Moments', *J. Op. Soc. of Amer.*, 70, Aug. 1980, pp. 920–930.

[1.86] L. A. Rubel, 'Internal and External Analytic Functions of Several Complex Variables', preprint provided to the author.

[1.87] Ph. Delsarte, Y. Genin and Y. Kamp, 'Planar Least-squares Inverse Polynomials Part 2: Asymptotic Behavior', *SIAM J. Alg. Disc. Meth.*, 1, Sept. 1980, pp. 336–344.

[1.88] E. Vanmarcke, *Random Fields: Analysis and Synthesis*, The MIT Press, Cambridge, MA, 1983.

[1.89] J. J. Rotmann, *An Introduction to Homological Algebra*, Academic Press, N.Y., 1979.

[1.90] I. A. Aizenberg and A. P. Yuzhakov, *Integral Representation and Residues in Multidimensional Complex Analysis*, Trans. of Math. Monographs, AMS, Vol. 58, Providence, R.I., 1983.

[1.91] N. K. Bose, Guest Editor, *Aspects of Spatial and Temporal Signal Processing*, Circuits, Systems and Signal Processing, Birkhäuser, Boston, Vol. 3, No. 2, 1984.

Chapter 2

N. K. Bose

Multivariate Rational Approximants of the Padé-Type in Systems Theory

2.1. INTRODUCTION AND MOTIVATION

While the study of Padé approximants is extensively documented in the literature [2.1], [2.2], its generalization to the multivariate case (henceforth to be referred to as Padé-type approximants) is of relatively recent origin. Nevertheless, though an early paper dealing with Padé-type approximants to a double power series was published as late as 1973, the ensuing decade witnessed a flurry of activities resulting in the documentation of those research not only in journals, but books [2.2, Section 1.4], [2.3, Section 4.2], special issues [2.4] and proceedings of conferences exclusively devoted to the topic of Padé approximation. Due to the very recent publication of the special issue on Rational Approximation for Systems, identified in reference [2.4], this chapter will be more concise than what was originally planned. The primary goal will be to alert the reader to the vast, albeit, scattered literature on multivariate rational approximants of the Padé type, and to acquaint him (her) with the possibility of applying these available resources to problems originating in systems theory. It is hoped that the reader will be able, without too much effort, to locate numerous other references, as needed, from the sources listed at the end of this chapter because these sources contain the necessary keys to those references, which, for practical limitations on space could not be included here.

Padé approximation theory has found numerous applications and interpretations in systems theory including the minimal partial realization problem [2.5], multiport network synthesis [2.6], time domain design of digital filters from a prescribed impulse response sequence [2.7], [2.8], and model reduction of control systems [2.9], [2.10]. The Padé-type of approximation theory also plays a significant role in multidimensional systems theory. In [2.11] algorithms for representation of double power series by branched continued fractions [2.12] has been considered with the objective of approximating 2–D input-output characteristics of nonlinear systems, described by a double power series. In [2.13], the problem

N. K. Bose (ed.), Multidimensional Systems Theory, 41–51
© 1985 *by D. Reidel Publishing Company.*

of order reduction of 2–D systems using Padé-type approximation has been considered. The approach is based on the matching of a finite number of coefficients in the double power series expansion of the transfer function of the system with the corresponding coefficients associated with the lower-order model, so that a number of linear equations in an equal number of unknowns are formed. No constraints, however, were placed on the geometry of the lattice points associated with the matched coefficients which increased the problem of computational complexity and, furthermore, an unstable lower-order model could result even though the original system transfer function was stable.

In the area of digital processing of 2–D signals, the problem of factoring spectral density functions occurs. It is well known that unlike in the 1–D case the 2–D spectral factorization of rational density functions into rational spectral factors is not possible, in general [2.14, Chapter 4]. However, it is possible to extend the concept of spectral factorization to the 2–D case provided non-rational infinite-order factors are allowed. Since, a recursive filter must be of finite order for implementation, it becomes necessary to obtain a rational approximation to the infinite-order factor. Padé-type approximation is possible in this case and, in order to attend to the problem of possible instability, it may be advisable to truncate as well as smooth the double power series characterizing the impulse response sequence (already obtained as an infinite-order spectral factor) by 2–D windows, as suggested in [2.15], prior to obtaining a desirable rational approximant. In the rational modeling of 2–D systems, the system to be modeled may not only be characterized by its impulse response sequence (first-order data) but also by its autocorrelation sequence (second-order data). In [2.16], a generalization of the Padé-type of approximation procedure has been considered to model a 2–D system characterized by both first- and second-order datas.

2.2. MULTIVARIATE PADÉ-TYPE APPROXIMANTS (SCALAR CASE)

For the sake of brevity in notation and exposition, we restrict ourselves to bivariate rational approximants of the Padé-type. It is noted that the generalization to the n-variate ($n > 2$) case is usually conceptually straightforward, even though it may be computationally quite combersome.

DEFINITION 2.1. A rational function, $[P(z_1, z_2)]/[Q(z_1, z_2)]$ is a bivariate approximant of the Padé-type to a double power series $T(z_1, z_2) = \Sigma_{i=0}^{\infty} \Sigma_{j=0}^{\infty} t_{ij} z_1^i z_2^j$ provided

$$T Q - P = \sum_{i=0}^{\infty} \sum_{j=0}^{\infty} r_{ij} z_1^i z_1^j \qquad (2.1)$$

with $r_{ij} = 0$ for $(i, j) \epsilon E$, where E denotes a suitable interpolation set. The choice of E and the set of points where the coefficients of $P(z_1, z_2)$ and $Q(z_1, z_2)$ are allowed to be nonzero determine the various types of approximants.

The Canterbury group's main efforts may be summarized by the following choices of polynomial coefficients support and interpolation set

$$P(z_1, z_2) = \sum_{i=0}^{n_1} \sum_{j=0}^{n_2} p_{ij} z_1^i z_2^j \qquad (2.2a)$$

$$Q(z_1, z_2) = \sum_{i=0}^{m_1} \sum_{j=0}^{m_2} q_{ij} z_1^i z_2^j \qquad (2.2b)$$

$$E = \{(i, j) |\ 0 \le i \le \max(n_1, m_1),\ 0 \le j \le \min(n_2, m_2)\}$$

$$U\{(i, j) |\ 0 \le i \le \min(n_1, m_1),\ 0 \le j \le \max(n_2, m_2)\}$$

$$U\{(i, j) |\ \max(n_2, m_2) < j \le n_2 + m_2,\ \max(n_2, m_2) <$$
$$< i + j \le n_2 + m_2,\ 0 \le i \le \min(n_1, m_1)\}$$

$$U\{(i, j) |\ \max(n_1, m_1) < i \le n_1 + m_1,\ \max(n_1, m_1) <$$
$$< i + j \le n_1 + m_1,\ 0 \le j \le \min(n_2, m_2)\} \qquad (2.2c)$$

with

$$r_{ij} = 0 \quad \text{for} \quad (i, j) \epsilon E \qquad (2.2d)$$

and (the symmetry conditions),

$$r_{m_1 + n_1 + 1 - \ell,\ \ell} + r_{\ell,\ m_2 + n_2 + 1 - \ell} = 0,$$
$$\ell = 1, 2, \ldots, \min(m_1, n_1, m_2, n_2). \qquad (2.2e)$$

Figure 2.1a shows the geometry of the lattice points forming the interpolation set E, described above. Chisholm [2.17] initiated the research of the Canterbury group by investigating approximants with $n_1 = n_2 = m_1 = m_2 = m$ in (2.2a)–(2.2e), which were shown to reduce to diagonal Padé approximants when one variable is equated to zero (projection),

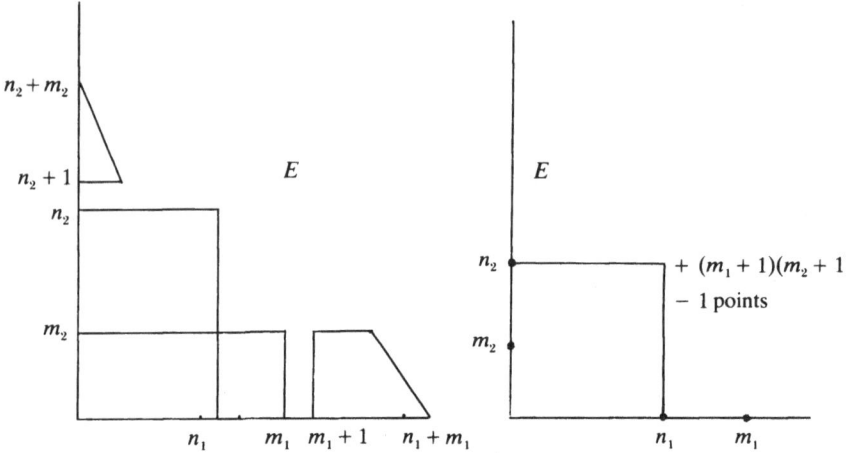

Fig 2.1(a) (Canterbury grid).   Fig. 2.1(b) (Lutterodt grid 1).

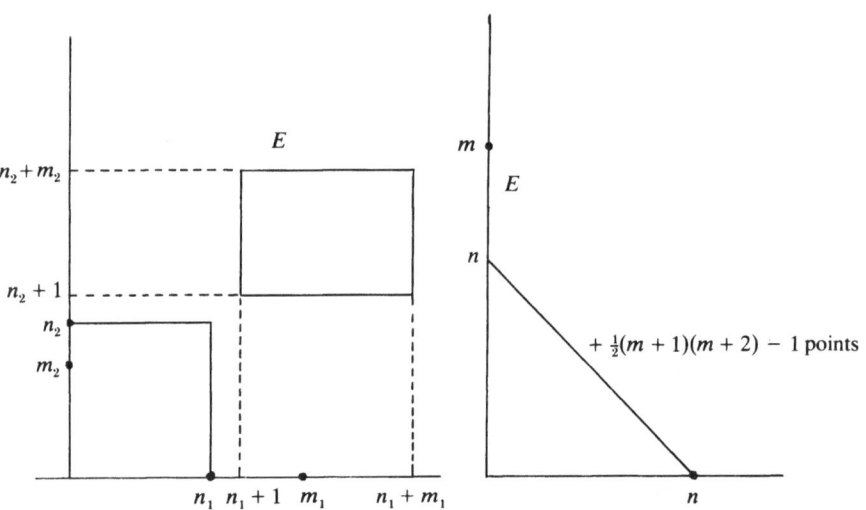

Fig. 2.1(c) (Lutterodt grid 2).   Fig. 2.1(d) (Karlsson-Wallin grid).

besides possessing properties of symmetry, existence and uniqueness, invariance under a group of certain bilinear transformations (homographic invariance), duality or reciprocal invariance, factorization and additivity. Chisholm and McEwan [2.18] provide the generalization of these results to the $n > 2$ case. Certain generalization of the bivariate and n-variate approximants in [2.17] and [2.18], respectively, were proposed [2.19], [2.20], as generalized analogues of off-diagonal Padé approximants. For determinantal expressions for the numerator and denominator polynomials of these approximants, see Levin [2.21].

Lutterodt [2.22], [2.23], considered approximants with $P(z_1, z_2)$ and $Q(z_1, z_2)$ defined as in (2.1a) and (2.1b), respectively, but his interpolating sets E were chosen as in Figures 2.1b and 2.1c. For Figure 2.1c [2.23],

$$
\begin{aligned}
E = \{(i, j)|\ 0 \le i \le n_1, 0 \le j \le n_2\} \\
U\{(i, j)|\ n_1 + 1 \le i \le n_1 + m_1, n_2 + 1 \le j \le n_2 + m_2\} \\
U\{(i, 0)|\ n_1 + 1 \le i \le n_1 + m_1\} \\
U\{(0, j)|\ n_2 + 1 \le j \le n_2 + m_2\}.
\end{aligned}
\tag{2.3}
$$

In both Figures (2.1b) and (2.1c) the set E has $(n_1 + 1)(n_2 + 1) + (m_1 + 1)(m_2 + 1) - 1$ elements which are required in the finding of (the same number of) unknown coefficients $\{p_{ij}\}, \{q_{ij}\}$ with $q_{00} \overset{\triangle}{=} 1$ by solving a set of linear equations derived from (2.1) by equating coefficients of monomials $z_1^i z_2^j$, $(i, j)\epsilon E$, on both sides of the equation. These approximants do not satisfy many of the properties of Chisholm's diagonal approximants. A detailed comparison of the properties of various types of approximants has been given by Cuyt [2.24]. Lutterodt considered the problem of generalizing his results to the n-variate case in [2.22].

Karlsson and Wallin considered Pade type rational approximants [2.25] where

$$
P(z_1, z_2) = \sum_{i+j=0}^{n} p_{ij} z_1^i z_2^j
\tag{2.4a}
$$

$$
Q(z_1, z_2) = \sum_{i+j=0}^{m} q_{ij} z_1^i z_2^j
\tag{2.4b}
$$

$$
E \supseteq \{(i, j): i + j \le n\}.
\tag{2.4c}
$$

The geometry of lattice points comprising set E, in this case, is shown in Figure 2.1d. Note that the number of unknowns in (2.4a) and (2.4b) with $q_{00} \triangleq 1$ is $\frac{1}{2}(n + 1)(n + 2) + \frac{1}{2}(m + 1)(m + 2) - 1$ which equals the number of elements in set E.

The computational complexity associated with the computation of the Padé-type of multivariate rational approximants is an important problem that deserves attention. In [2.26] and in [2.19], a computational scheme, called the prong method, has been used to obtain a solution to the Canterbury approximants and numerical difficulties encountered in the construction of such approximants were discussed. In many situations, it is advisable from the practical standpoint, that the approximation problem be so formulated such that recursive schemes for computation may be developed by exploiting certain structural properties of the characterizing matrix in the system of linear equations, which is set up for solution. The recursive scheme may also be developed as in the univariate case by applying the so called ϵ-algorithm due to Wynn [2.3, pp. 159–177]. It is well-known that if this ϵ-algorithm is applied to the partial sums of the univariate power series, $T(z) = \Sigma_{i=0}^{\infty} t_i z^i$, then $\epsilon_{2m}^{(1-m)}$ (this is standard notation for the ϵ-algorithm) is the $[1/m]$ Padé approximant to $T(z)$ i.e. $[1/m]$ has polynomials of degrees 1 and m, respectively, for its numerator and denominator. Hillion [2.27] used the ϵ-algorithm for obtaining rational approximations for multiple power series. Cuyt [2.28] applied the ϵ-algorithm for recursive computation of multivariate Padé-type approximants introduced by her [2.29] and proved that these approximants satisfy the property analogous to the one, referred to above, which is valid for Padé approximants (univariate case). In [2.30], it was shown how the 1–D Padé technique may be applied several times to obtain a bivariate rational approximant with the power series coefficient matching property (i.e. matched with a finite set of coefficients of a specified power series) over a rectangular grid. The technique generalizes naturally to the n–D, $n > 2$, case, since the generalization calls for the repeated application of the 1–D Padé approximation technique to a univariate power series expansion whose coefficients belong either to the real number field or the field of rational functions in one or more independent complex variables. Though the technique is conceptually simple and appears to be computationally expedient, the properties of the rational approximants generated remain to be studied in depth.

2.3. PADE-TYPE MATRIX APPROXIMANTS

Padé approximants to an operator $T(z)$, analytic in z and possessing a formal Taylor series expansion around $z = 0$ with coefficients in a non-commutative algebra, were considered by Bessis in 1973 [2.31], and the various invariance properties of such approximants were also studied. The feasibility of recursive computation of 1–D matrix Padé approximants was demonstrated in [2.32]. Matrix Stieltjes series was defined in [2.6], and the RC-ideal transformer network realizability property for matrix Padé approximants of certain orders to such series was proved via the artifices of continued-fraction expansion and matrix Cauchy index. A multivariate scalar Stieltjes series can be defined. For the sake of brevity a 2–D Stieltjes series is defined next.

DEFINITION 2.2. Let,

$$t_{ij} = \int_0^\infty \int_0^\infty x_1^i x_2^j \, d\sigma(x_1, x_2)$$

be finite, $\forall i, j$, where $\sigma(x_1, x_2)$ on $0 \le x_1 < \infty$, $0 \le x_2 < \infty$ is bounded, monotone nondecreasing in x_1 for fixed x_2, and monotone nondecreasing in x_2 for fixed x_1. Then, the series

$$T(z_1, z_2) = \sum_{i=0}^\infty \sum_{j=0}^\infty t_{ij} \, z_1^i \, z_2^j$$

is a double Stieltjes series with an integral representation,

$$T(z_1, z_2) = \int_0^\infty \int_0^\infty \frac{d\sigma(x_1, x_2)}{(1 - z_1 x_1)(1 - z_2 x_2)}.$$

Actually, the integral representation in Definition 2.2 may be considered as one possible generalization of a Stieltjes function from the 1–D to the 2–D case. Another possible representation for a bivariate Stieltjes function is,

$$T(z_1, z_2) = \int_0^\infty \int_0^\infty \frac{d\sigma(x_1, x_2)}{(1 - z_1 x_1 - z_2 x_2)}.$$

whose power series expansion is [2.33],

$$T(z_1, z_2) = \sum_{i=0}^\infty \sum_{j=0}^\infty \binom{i+j}{j} t_{ij} \, z_1^i \, z_2^j$$

where $t_{ij's}$ are the moments defined in Definition 2.2. It will be useful to define a bivariate matrix Stieltjes series with the objective of relating Padé-type approximants of specific orders to such series with the impedance matrices of a subclass of passive multivariate multiports [2.14, Chapter 5]. In fact, this study may be initiated with the scalar bivariate Stieltjes series and one-ports with driving-point impedances,

$$Z(z_1, z_2) = F + \frac{T_1}{z_1} + \frac{T_2}{z_2}$$

where F, T_1, T_2, are real nonnegative functions in the complex variables z_1, z_2 and $Z(z_1, z_2)$ is a real rational function in z_1, z_2. In the 1–D case, Padé approximants of certain orders to a Stieltjes series are impedance functions of the form

$$Z(z_1) = F + \frac{T_1}{z_1}$$

whose poles and zeros are simple and alternate on the negative real axis with the pole being closest to the origin. Studies relating matrix Stieltjes series to multivariate multiport realizability theory will be useful in the derivation of new properties of Padé-type approximants to matrix Stieltjes series in addition to those which are known [2.33].

The bivariate Padé-type approximation to a general bivariate matrix power series has been considered in [2.34], where recursive schemes for computing the approximants have been developed. Since these results have potential applications in multivariable (multi-input/multi-output) 2–D system design, a sufficient condition for testing the stability of the approximants in terms of the specified impulse response sequence has been obtained [2.35].

2.4. CONCLUSIONS

This chapter provides a concise exposition of the status of multivariate rational approximation theory of the Padé type and its links to multidimensional systems theory. It is, however, emphasized, that problems of interest in multidimensional systems theory like the identification of linear systems, stabilization of spatio-temporal and distributed recursive filters and the modeling of random fields by autoregressive moving-average processes might not only be based on either the specified impulse-response or autocorrelation data but on both. When only the first order (impulse response) data is provided (in, for instance, modeling from a deterministic point of view) one might proceed with Padé type of approximants while when only the second order datas are provided (modeling from a stochastic point of view by fitting the system's input-output covariance properties), the least squares approximation theory is widely used. In the later case, for 2–D problems the normal equations involve a positive definite block Toeplitz matrix, and 2–D Levinson type algorithm is available [2.36] for fast computation of the solution vector. There is a significant body of current literature available on approximants based on the least-squares criterion [2.14, Chapter 4], [2.37], but the objective here is not to describe that aspect of research. Also, the use of a modified least-squares procedure to obtain approximants using first and second order datas was cited [2.16].

Several problems remain to be investigated. First, the availability of a three term recurrence formula relating the denominator polynomials of bivariate approximants of successive orders [2.34], suggests that these polynomials may belong to a sequence of bivariate polynomials which are orthogonal over real intervals [2.38]. This possibility has to be checked for validity. Second, it is of mathematical interest to develop a theory of bivariate matrix orthogonal polynomials over real intervals analogous to what is recently available in the single variable case [2.6], [2.39]. The 'denominator' bivariate matrix polynomials in the matrix rational approximants might form such an orthogonal sequence. Third, as described in Section 2.3, a study of multivariate matrix Stieltjes series would be useful, especially in the context of its possible relation to multivariate network realizability theory. Fourth, in the context of 2–D digital filtering, it is useful to determine whether or not an approximation scheme similar to that in [2.34] can be developed for the design of half-plane filters [2.14, Chapter 3]. It is expected that a judicious blend of techniques used in approximating a formal Laurent series by means of rational functions [2.40] with the techniques used in [2.34] might help in reaching this goal.

REFERENCES

[2.1]   G. E. Baker Jr. and Peter Graves-Morris, *Padé Approximants Part 1: Basic Theory*, *Encyclopedia of Mathematics and Its Applications*, Vol. 13, Addison-Wesley Publ. Co., Reading, MA, 1981.

[2.2] G. E. Baker Jr. and Peter Graves-Morris, *Padé Approximants Part 2: Extensions and Applications, Encyclopedia of Mathematics and Its Applications*, Vol. 14, Addison-Wesley Publ. Co., Reading, MA, 1981.

[2.3] C. Brezinski, *Padé-Type Approximation and General Orthogonal Polynomials*, ISNM50, Birkhäuser Verlag, Basel, 1980. (Section 4.2).

[2.4] J. S. R. Chisholm, 'Generalizations of Padé Approximants', in special issue: *Rational Approximations for Systems*, A. Bultheel and P. Dewilde (eds.), Circuits, Systems and Signal Processing, Birkhäuser Boston Inc., 1, Nos. 3–4, 1982, pp. 279–287.

[2.5] B. W. Dickinson, M. Morf, and T. Kailath, 'A Minimal Realization Algorithm for Matrix Sequences', *IEEE Trans. Auto. Control*, 10, Feb. 1974, pp. 31–38.

[2.6] S. Basu and N. K. Bose, 'Matrix Stieltjes Series and Network Models', *SIAM J. Math. Anal.*, 14, March 1983, pp. 209–222.

[2.7] R. Hastings James and S. K. Mehra, 'Extensions of the Padé-approximation Technique for the Design of Recursive Digital Filters', *IEEE Trans. ASSP*, 25, Dec. 1977, pp. 501–509.

[2.8] C. S. Burrus and W. Parks, 'Time Domain Design of Recursive Digital Filters', *IEEE Trans. Audio and Electroacoustics*, 18, June 1970, pp. 137–141.

[2.9] Y. Shamash, 'Stable Reduced Order Model Using Padé-Type Approximation', *IEEE Trans. Auto. Control*, 19, Oct. 1974, pp. 615–616.

[2.10] Y. Shamash, 'Model Reduction Using the Routh Stability Criterion and the Padé Approximation Technique', *Int. J. Control*, 21, 1975, pp. 475–484.

[2.11] A. Cichocki, 'Nested-feedback-loops Realization of 2–D Systems', *Circuits, Systems and Signal Proc.*, Birkhäuser Boston Inc., 1, Nos. 3–4, 1982, pp. 321–343.

[2.12] W. Siemaszko, 'Branched Continued Fraction for Double Power Series', *J. Comp. Appl. Math.*, No. 2, 1980, pp. 121–125.

[2.13] P. N. Paraskevopoulos, 'Padé-type Order Reduction of Two-dimensional Systems', *IEEE Trans. Circuits and Systems*, 27, May 1980, pp. 413–416.

[2.14] N. K. Bose, *Applied Multidimensional Systems Theory*, Van Nostrand Reinhold Co., New York, 1982.

[2.15] M. P. Ekstrom and J. W. Woods, 'Two-dimensional Spectral Factorization with Applications in Recursive Digital Filtering', *IEEE Trans. ASSP*, 24, April 1976, pp. 115–128.

[2.16] L. F. Chaparro and E. I. Jury, 'Rational Approximation of 2–D Linear Discrete Systems', *IEEE Trans. ASSP*, 30, Oct. 1982, pp. 780–787.

[2.17] J. S. R. Chisholm, 'Rational Approximants Defined from Double Power Series', *Mathematics of Computation*, 27, No. 124, Oct. 1973, pp. 841–848.

[2.18] J. S. R. Chisholm and J. McEwan, 'Rational Approximants Defined from Power Series in n-Variables', *Proc. R. Soc. Lond.* A.336, 1974, pp. 421–452.

[2.19] R. Hughes Jones and G. J. Makinson, 'The Generation of Chisholm Rational Polynomial Approximants in Power Series in Two Variables', *J. Inst. Maths. Applics.*, 13, 1974, pp. 299–310.

[2.20] R. Hughes Jones, 'General Rational Approximants in n-Variables', *Jour. Approx. Theory*, 16, 1976, pp. 201–233.

[2.21] D. Levin, 'General Order Padé-type Rational Approximants Defined from Double Power Series', *J. Inst. Maths. Applics*, 18, 1976, pp. 1–8.

[2.22] C. H. Lutterodt, 'Rational Approximants to Holomorphic Functions in n-Dimensions', *Jour. Mathematical Analysis and Applications*, 53, 1976, pp. 89–98.

[2.23] C. H. Lutterodt, 'A Two-dimensional Analogue of Padé Approximant Theory', *J. Phys. Math.*, 7, 1974, pp. 1027–1037.

[2.24] Annie A. M. Cuyt, 'Abstract Padé Approximants for Operators: Theory and Applications', Doctoral Thesis, Universiteit Antwerpen, Department Wiskunde, Antwerp, Belgium, 1982.

[2.25] J. Karlsson and H. Wallin, 'Rational Approximation by an Interpolation Procedure in Several Variables', in *Padé and Rational Approximation: Theory and Applications*, E. B. Saff and R. S. Varga (eds.), Academic Press, London 1977.

[2.26] P. R. Graves-Morris, R. Hughes Jones and G. J. Makinson, 'The Calculation of Rational Approximants in Two Variables', *J. Inst. Maths. Applics.*, 13, 1974, pp. 299–310.

[2.27] P. Hillion, 'Remarks on Rational Approximations of Multiple Power Series', *J. Inst. Math. Applics.*, 19, 1977, pp. 281–293.

[2.28] Annie A. M. Cuyt, 'The ϵ-Algorithm and Multivariate Padé Approximants', *Numerische Mathematik*, 40, 1982, pp. 39–46.

[2.29] Annie A. M. Cuyt, 'Abstract Padé Approximants in Operator Theory', in *Padé Approximation and its Applications*, L. Wuytack (ed.), Lecture Notes in Mathematics, LNM 765, Springer-Verlag, Berlin, pp. 61–87.

[2.30] N. K. Bose, 'Two-dimensional Rational Approximants via One-dimensional Padé Technique', *Proc. of European Signal Proc. Conf.*, (Kunt and deCoulon (eds.)), Lausanne, Switerland, 1980, pp. 409–411.

[2.31] D. Bessis, 'Topics in the Theory of Padé Approximants', in *Padé Approximants*, P. R. Graves-Morris (ed.), The Institute of Physics, London and Bristol, 1973, pp. 20–44.

[2.32] N. K. Bose and S. Basu, 'Theory and recursive Computation of 1–D Matrix Padé Approximants', *IEEE Trans. Circuits and Systems*, 27, April 1980, pp. 323–325.

[2.33] M. F. Barnsley and P. D. Robinson, 'Rational Approximant Bounds for a Class of Two-variable Stieltjes Functions', *SIAM J. Math. Anal.*, 9, April 1978, pp. 272–290.

[2.34] N. K. Bose and S. Basu, 'Two-dimensional Matrix Padé Approximants: Existence, Nonuniqueness, and Recursive Computation', *IEEE Trans. Auto. Control*, 25, June 1980, pp. 509–514.

[2.35] S. Basu and N. K. Bose, 'Stability of 2–D Matrix Rational Approximants from Input Data', *IEEE Trans. Auto. Control*, 26, April 1981, pp. 540–541.

[2.36] J. H. Justice, 'A Levinson-type Algorithm for Two-dimensional Wiener Filtering Using Bivariate Szego Polynomials', *Proc. of IEEE*, 65, June 1977, pp. 582–586.

[2.37] T. L. Marzetta, 'Two-dimensional Linear Prediction: Autocorrelation Arrays, Minimum-phase Prediction Error Filters and Reflection Coefficient Arrays', *IEEE Trans. ASSP*, 28, Dec. 1980, pp. 725–733.

[2.38] D. Jackson, 'Formal Properties of Orthogonal Polynomials in Two Variables', *Duke Math. J.*, 2, 1936, pp. 423–434.

[2.39] J. S. Geronimo, 'Scattering Theory and Matrix Orthogonal Polynomials on the Real Line', *Circuits, Systems, and Signal Processing*, 1, Nos. 3–4, 1982, pp. 471–495.

[2.40] W. B. Gragg and G. D. Johnson, 'The Laurent-Padé Table', *Information Processing 74*, Proc. IFIP. Congress 74, North-Holland, Amsterdam, 1974, pp. 632–637.

Chapter 3

J. P. Guiver and N. K. Bose

Causal and Weakly Causal 2–D Filters with Applications in Stabilization

3.1. SCALAR 2–D INPUT/OUTPUT SYSTEMS

A discrete 2–D scalar input/output system is described by means of an equation

$$\{y(m, n)\} = F(\{x(m, n)\}) \tag{3.1}$$

where F is an operator which maps doubly indexed input sequences $\{x(m, n)\}$ into doubly indexed output sequences $\{y(m, n)\}$ where we assume that $x(m, n)$ and $y(m, n)$ take values in $\mathbb{R}$, the field of real numbers.

An important class of systems are the linear shift-invariant (LSI) system. The input/output relation of an LSI system can be represented by means of a convolution equation

$$y(m, n) = \sum_{(k, \ell) \in \mathbb{Z}^2} h(m - k, n - \ell) x(k, \ell) \tag{3.2}$$

where

$$h(m, n) = F(\delta(m, n)) \tag{3.3}$$

is the system response to a unit impulse $\delta(m, n)$ applied at $(0, 0)$, and $Z^2 = Z \times Z$, where Z is the set of integers.

We will denote the support of the sequence $\{h(m, n)\}$ by $\mathrm{Supp}(h)$, i.e.

$$\mathrm{Supp}(h) = \{(r, s) \in \mathbb{Z}^2 : h(r, s) \neq 0\}. \tag{3.4}$$

Usually some sort of causality is assumed which is specified by conditions on $\mathrm{Supp}(h)$.

(3.5). DEFINITION. We will say an LSI system F is *causal* if

$$\mathrm{Supp}(h) \subset Q_1 \text{ where } Q_1 = \{(x, y) \in \mathbb{R}^2 : x \geq 0, y \geq 0\}. \tag{3.6}$$

We will say F is *weakly causal* if (defined by R. Eising in [3.1])

$$\mathrm{Supp}(h) \subset C \tag{3.7}$$

N. K. Bose (ed.), Multidimensional Systems Theory, 52–100

where C is a closed convex cone in $\mathbb{R}^2$ satisfying

(i) $C \cap (-C) = \{(0, 0)\}$

(ii) $Q_1 \subset C$. $\qquad\qquad\qquad\qquad\qquad$ (3.8)

The first condition merely says that the cone makes an angle of less than π at $(0, 0)$.

Under condition (3.7) any input $x(m, n)$ which satisfies

$$\text{Supp}(u_{mn}) \cap C \text{ is finite for all } (m, n)\epsilon\mathbb{Z}^2 \qquad (3.9)$$

where

$$u_{mn}(r, s) \overset{\Delta}{=} x(m - r, n - s)$$

will give rise to a unique output by means of the convolution equation (3.2).

If $a(m, n)$ is any doubly indexed sequence, we define its 2–D Z-transform by

$$Z(a(m, n)) = A(z, w) = \sum_{(m, n)\epsilon\mathbb{Z}^2} a(m, n)z^m w^n. \qquad (3.10)$$

If we then take the Z-transform of each side of (3.2) we get

$$Y(z, w) = H(z, w)X(z, w). \qquad (3.11)$$

(3.12). DEFINITION. $H(z, w) = \sum_{(m, n)\epsilon\mathbb{Z}^2} h(m, n)z^m w^n$ is called the system function.

$H(z, w)$ can be regarded either as a formal power series or as a function of z and w when the power series converges.

From a practical standpoint it is desirable to approximate $H(z, w)$ by means of a rational function in z, w, z^{-1} and w^{-1} for purposes of recursive implementation of the input/output relation.

Let

$$a(z, w) = \sum_{(m, n)\epsilon A} \alpha(m, n)z^m w^n$$

where $A \subset \mathbb{Z}^2$ is finite and

$$(0, 0)\epsilon A \quad \text{and} \quad \alpha(0, 0) = 1. \qquad (3.13)$$

Then for recursive implementation of $H(z, w) = [b(z, w)]/[a(z, w)]$ by means of a 2–D difference equation it is required (see O'Connor and

Huang [3.2]) that A be contained in a lattice sector with vertex $(0, 0)$ of angle less than π.

Here $b(z, w) = \Sigma_{(m, n)\epsilon B} \beta(m, n)z^m w^n$ where it is assumed that $B \subset \mathbb{Z}^2$ is finite, but, for recursive computability, no other conditions are needed on B.

However, we will assume that

$$A, B \subset C \tag{3.14}$$

where C is a closed convex cone satisfying (3.8). We then have the following observation.

(3.15). PROPOSITION. *Let* $a(z, w)$, $b(z, w)\epsilon R[z, z^{-1}, w, w^{-1}]$ *with supports* A, B *respectively where* A *satisfies* (3.13) *and* A *and* B *satisfy* (3.14). *Then* $[b(z, w)]/[a(z, w)]$ *has a unique expansion* $\Sigma h(m, n)z^m w^n$ *such that* $\text{Supp}(h) \subset C$.

Proof. The existence of the expansion follows since

$$\frac{b(z, w)}{a(z, w)} = \frac{b(z, w)}{1 - a^*(z, w)}$$

$$= b(z, w) \sum_{i = 0}^{\infty} [a^*(z, w)]^i$$

where

$$a^* = - \sum_{A - \{(0, 0)\}} a(m, n)z^m w^n.$$

It is then clear that all exponents must be in C.

The uniqueness follows from the work of Eising ([3.1], Lemma 2–10), which we will discuss later in this chapter.

Eising [3.3] shows that a causal impulse response has a finite dimensional local state space realization if (and clearly only if) $H(z, w)$ has a representation of the form $[p(z, w)]/[q(z, w)]$ where $p, q\epsilon\mathbb{R}[z, w]$ and $q(0, 0) \neq 0$.

(3.16). DEFINITION. A rational function $[p(z, w)]/[q(z, w)]$ with $p, q\epsilon\mathbb{R}[z, w]$ is called *causal* if $q(0, 0) \neq 0$. It is called *strictly causal* if in addition $p(0, 0) = 0$.

It follows from (3.15) that this definition is consistent with the definition in (3.6). Strict causality will imply that $h(0, 0) = 0$. i.e. the output will not depend on the current input.

Eising [3.1] also shows that if $H(z, w)$ has a representation of the form $[b(z, w)]/[a(z, w)]$ where $a, b \in \mathbb{R}[z, z^{-1}, w, w^{-1}]$ and a, b have supports A, B respectively, satisfying (3.13) and (3.14), then $H(z, w)$ has a finite dimensional local state space realization.

(3.17). DEFINITION. A function of the form $[b(z, w)]/[a(z, w)]$ where a and b are as described in the above paragraph is called *weakly causal*. It is called *strictly weakly causal* if in addition $b(z, w)$ has zero constant term.

This definition is, from (3.15) seen to be consistent with (3.7). Strictly weakly causal will imply that the output will not depend on the current input – i.e. there is no direct coupling.

The systems we consider in this chapter will be of the type described in (3.16) and (3.17) – i.e. (strictly) causal and (strictly) weakly causal rational transfer functions.

3.2. STABILITY

(3.18). A I/O system (3.1) is said to to BIBO (bounded input/bounded output) stable if for all $M > 0$ there exists $N > 0$ such that for any input sequence $\{x\}$ satisfying $|x(m, n)| < M$ for all (m, n), the corresponding output sequence satisfies $|y(m, n)| < N$ for all (m, n).

For a system described by a convolution equation (3.2), BIBO stability is well known to be equivalent to

$$\sum_{m, n} |h(m, n)| < \infty. \tag{3.19}$$

Given a cone C satisfying (3.8), $C \subset \mathbb{Z}^2$ is contained in a lattice sector (with vertex angles less than π) of the form $S[(M_1, N_1), (M_2, N_2)]$ defined by the relatively prime integer pairs (M_1, N_1) and (M_2, N_2) (see [3.2]). We assume that (M_1, N_1) and (M_2, N_2) are independent vectors; otherwise, we can map $S[(M_1, N_1), (M_2, N_2)]$ on to $\mathbb{Z}_+$ and we would thus be dealing with an essentially 1–D situation (see [3.2]).

In general $S[(M_1, N_1), (M_2, N_2)]$ can be mapped into (but not necessarily onto) $Q_1 \cap \mathbb{Z}^2$ by means of the map

$$\begin{aligned} m' &= k_1 m + k_2 n \\ n' &= k_3 m + k_4 n \end{aligned} \tag{3.20}$$

where

$$k_1 = \text{sgn}(D)N_2$$
$$k_2 = -\text{sgn}(D)M_2$$
$$k_3 = -\text{sgn}(D)N_1 \qquad (3.21)$$
$$k_4 = \text{sgn}(D)M_1$$

and

$$D = M_1N_2 - M_2N_1. \qquad (3.22)$$

(Note: $D \neq 0$ by the assumption that (M_1, N_1) and (M_2, N_2) be independent). Then we have the following theorem of O'Connor and Huang [3.2].

(3.23). THEOREM. Let $\text{Supp}(h) \subset S[(M_1, N_1), (M_2, N_2)]$ (with vertex angle less than π). Then $\Sigma h(m, n)z^m w^n$ is BIBO stable if and only if $\Sigma g(m', n')z^{m'} w^{n'}$ is BIBO stable where

$$g(m', n') = g(k_1m + k_2n, k_3m + k_4n) = h(m, n). \qquad (3.24)$$

(3.23) says that we can determine the stability of a weakly causal transfer function by checking that of an appropriate causal one.

As noted earlier, the map given in (3.20) does not in general map onto the whole of $Q_1 \cap \mathbb{Z}^2$. Eising ([3.1], Lemma 2–5) remedied this by showing that any cone C satisfying (3.8) is contained in a causality cone C_c. $C_c \cap \mathbb{Z}^2$ can then be mapped bijectively (i.e. in a 1–1, onto manner) to $Q_1 \cap \mathbb{Z}^2$. For convenience we reproduce (but don't prove) some details of Eising's work. (See [3.1] for full details).

(3.25). DEFINITION. A causality cone C_c is an intersection of two half-planes $H_{p, r}$ and $H_{q, t}$ where

$$H_{p, r} = \{(x, y): (x, y)\epsilon\mathbb{R}^2, px + ry \geq 0\}$$
$$H_{q, t} = \{(x, y): (x, y)\epsilon\mathbb{R}^2, qx + ty \geq 0\} \qquad (3.26)$$

where p, q, r, t are non-negative integers satisfying

$$pt - qr = 1. \qquad (3.27)$$

Note that $(H_{p, r} \cap H_{q, t}) \cap \mathbb{Z}^2 = S[(t, -q), (-r, p)]$.

(3.28). LEMMA. Let C be a closed convex cone satisfying (3.8). Then there exists a causality cone C_c such that $C \subset C_c$.

(3.29). Let $C_c = H_{p,r} \cap H_{q,t}$ be a causality cone then the map ϕ: $C_c \cap \mathbb{Z}^2 \to Q_1 \cap \mathbb{Z}^2$ given by

$$\phi(m, n) = (pm + rn, qm + tn) \tag{3.30}$$

is bijective.

(ϕ is the same map as described in (3.20) and (3.21) with $M_1 = t$, $N_1 = -q$, $M_2 = -r$, $N_2 = p$ and $D = pt - qr = 1$).

It is important to note that for $C_c = H_{p,r} \cap H_{q,t}$, $C_c \cap \mathbb{Z}^2$ is a semi-group under addition, with $(0, 0)$ as 'identity', which is generated by $(t, -q)$ and $(-r, p)$. (In general a lattice sector $S[(M_1, N_1), (M_2, N_2)]$ is a semigroup but is not generated by (M_1, N_1) and (M_2, N_2), or by any two vectors). The fact that $(t, -q)$ and $(-r, p)$ generate $C_c \cap \mathbb{Z}^2$ follows from the fact that the ϕ given above is a semigroup isomorphism which maps $(t, -q)$ to $(1, 0)$ and $(-r, p)$ to $(0, 1)$. Given $(m, n) \epsilon C_c \cap \mathbb{Z}^2$, $\phi(m, n) = (h, k) = h(1, 0) + k(0, 1)$.

Therefore

$$(m, n) = h(t, -q) + k(-r, p)$$
$$= (ht - kr, -hq + kp)$$

where

$$h = pm + rn \qquad k = qm + tn.$$

(3.31). NOTATION. For any $C_c = H_{p,r} \cap H_{q,t}$ we let $S_{p,r,q,t} = \{H(z, w): \exists h(m, n) \text{ such that } H(z, w) = \Sigma h(m, n)z^m w^n \text{ and Supp}(h) \subset C_c\}$

With this notation the ring of causal power series is represented by $S_{1,0,0,1}$ since $Q_1 = H_{1,0} \cap H_{0,1}$.

THEOREM. $S_{p,r,q,t}$ is a ring which is isomorphic to $S_{1,0,0,1}$. The ring isomorphism

$$\Phi: S_{p,r,q,t} \to S_{1,0,0,1} \tag{3.32}$$

is given by

$$\Phi(H)(\alpha, \beta) = \sum_{m=0}^{\infty} \sum_{n=0}^{\infty} h(\phi^{-1}(m, n))\alpha^m \beta^n \tag{3.33}$$

where ϕ is as in (3.30).

(3.34). LEMMA. The isomorphism Φ can be described by the substitution

$$z = \alpha^p \beta^q \qquad w = \alpha^r \beta^t \tag{3.35}$$

with inverse

$$\alpha = z^t w^{-q} \qquad \beta = z^{-r} w^p. \tag{3.36}$$

The map Φ is of course different for different values of p, r, q, t but use of the same notation should not cause any confusion.

(3.37). NOTATION. For any $C_c = H_{p,r} \cap H_{q,t}$ we let $P_{p,r,q,t}$ represent the ring of sums of the form $\Sigma_{(m,n)\in A}\, \alpha(m,n)z^m w^n$ where A is finite and $A \subset C_c$. Thus $P_{1,0,0,1}$ is just $\mathbb{R}[z, w]$, the ring of two variable polynomials over R.

(3.38). NOTATION. For $C_c = H_{p,r} \cap H_{q,t}$ we let $R_{p,r,q,t}$ denote the subring of $S_{p,r,q,t}$ consisting of transfer functions representable in the form $[b(z, w)]/[a(z, w)]$ where a, $b, \in P_{p,r,q,t}$ and $a(z, w)$ has non-zero constant term.

(3.39). We let $\bar{R}_{p,r,q,t}$ denote the subring of $R_{p,r,q,t}$ consisting of those functions for which the numerator $b(z, w)$ $(\in P_{p,r,q,t})$ has constant term equal to zero.

Thus, in the above notations, $R_{1,0,0,1}$ and $\bar{R}_{1,0,0,1}$ stand for the rings of causal and strictly causal rational functions respectively.

It follows from (3.35) and (3.36) that the Φ given in (3.33) maps

$$P_{p,r,q,t} \text{ bijectively onto } P_{1,0,0,1}$$
$$R_{p,r,q,t} \text{ bijectively onto } R_{1,0,0,1} \tag{3.40}$$

and

$$\bar{R}_{p,r,q,t} \text{ bijectively onto } \bar{R}_{1,0,0,1}.$$

(3.41). Given $a(z, w)\in P_{p,r,q,t}$ it follows from the observations following (3.30) that $a(z, w)$ can be written uniquely as a polynomial in $z^t w^{-q}$ and $z^{-r} w^p$ and then Φ is just that map (see (3.35) and (3.36)) which replaces $z^t w^{-q}$ by α and $z^{-r} w^p$ by β.

Similar comments apply for $R_{p,r,q,t}$ and $\bar{R}_{p,r,q,t}$ which are just quotients of elements in $P_{p,r,q,t}$.

(3.42) DEFINITIONS. A representation (b/a) $(a, b, \epsilon P_{p, r, q, t}$ with a having non-zero constant term) of an element in $R_{p, r, q, t}$ will be called irreducible if there does not exist $a_1, b_1, c\epsilon P_{p, r, q, t}$, with $c(z, w)$ not a constant, such that $a = a_1 c$, $b = b_1 c$; in other words if $a(z, w)$ and $b(z, w)$ have no non-trivial common factor in the ring $P_{p, r, q, t}$.

Now, in general, given two arbitrary elements, $a(z, w)$ and $b(z, w)$, in $P_{p, r, q, t} \cap P_{p', r', q', t'}$ with no restriction on $a(z, w)$ we can have a and b being relatively prime in one ring but not in the other. For example, $b(z, w) = z$, $a(z, w) = z + w$ are relatively prime in $P_{1, 0, 0, 1}$ but have w as a common factor when considered as elements of $P_{1, 0, 1, 1}$. However, the following proposition is true.

(3.43). PROPOSITION. *Let, $a, b\epsilon P_{p, r, q, t} \cap P_{p', r', q', t'}$.*

Then if a and b have no (non-trivial) common factor in $P_{p, r, q, t}$, $a(z, w)$ and $b(z, w)$ can only have common factors of the form (const.)$z^k w^\ell$ $((k, \ell)\epsilon H_{p', r'} \cap H_{, q', t'})$ in $P_{p', r', q', t'}$.

Proof. Suppose $a(z, w)$ and $b(z, w)$ have a (non-constant) common factor $c(z, w)\epsilon P_{p', r', q', t'}$ which is not of the form (const.) $z^k w^\ell$ $((k, \ell) \epsilon H_{p', r'} \cap H_{q', t'})$. Then there exist $a_1(z, w), b_1(z, w)\epsilon P_{p', r', q', t'}$ such that

$$a(z, w) = a_1(z, w)\, c\, (z, w)$$
$$b(z, w) = b_1(z, w)\, c\, (z, w). \tag{3.44}$$

After multiplying each side of (3.44) by appropriate powers of z and w we see that there exists $m, n \geq 0$ such that

$$z^m w^n a(z, w) \quad \text{and} \quad z^m w^n b(z, w) \tag{3.45}$$

are in $\mathbb{R}[z, w]$ and have a common factor in $\mathbb{R}[z, w]$ not of the form (const.) $z^\mu w^\nu$, $\mu, \nu \geq 0$.

Let

$$\Phi: P_{p, r, q, t} \to \mathbb{R}[z, w] \qquad (\text{see } (3.40))$$

take

$$a(z, w) \quad \text{to} \quad a_0(\alpha, \beta)$$

and

$$b(z, w) \quad \text{to} \quad b_0(\alpha, \beta)$$

Then Φ takes (see (3.35))

$$z^m w^n a(z, w) \quad \text{to} \quad \alpha^{pm + rn}\beta^{qm + tn}a_0(\alpha, \beta)$$

and

$$z^m w^n b(z, w) \quad \text{to} \quad \alpha^{pm + rn}\beta^{qm + tn}b_0(\alpha, \beta)$$

Now (3.45) indicates that $\alpha^{pm + rn}\beta^{qm + tn}a_0(\alpha, \beta)$ and $\alpha^{pm + rn}\beta^{qm + tn}b_0(\alpha, \beta)$ have a common factor not of the form (const.) $\alpha^i\beta^j$, $i, j \geq 0$. Therefore, $a_0(\alpha, \beta)$ and $b_0(\alpha, \beta)$ must have such a common factor, say $d(\alpha, \beta)$.

Then $d(z^t w^{-q}, z^{-r}w^p)$ is a non-trivial factor of

$$a(z, w) = a_0(z^t w^{-q}, z^{-r}w^p)$$

and

$$b(z, w) = b_0(z^t w^{-q}, z^{-r}w^p)$$

contradicting the hypothesis that $a(z, w)$ and $b(z, w)$ are relatively prime in $P_{p, r, q, t}$.

(3.46). COROLLARY. *Let* $\quad p(z, w) = [b(z, w)]/[a(z, w)]\epsilon R_{p, r, q, t} \cap R_{p', r', q', t'}$. *(So a, $b\epsilon P_{p, r, q, t} \cap P_{p', r', q', t'}$ and $a(z, w)$ has a non-zero constant term). Then $[b(z, w)]/[a(z, w)]$ is irreducible in $R_{p, r, q, t}$ if and only if it is irreducible in $R_{p', r', q', t'}$.*

Proof. Since $a(z, w) = \text{const.} + \text{other terms}$, a factor of the form $z^k w^\ell$ would mean $a(z, w) = z^k w^\ell$ ((const.) $z^{-k}w^{-\ell} + \text{other terms}$). But (k, ℓ) and $(-k, -\ell)$ cannot lie in the same causality cone unless $k = \ell = 0$. So $a(z, w)$ cannot have any non-trivial factors of the form $z^k w^\ell$ either in $P_{p, r, q, t}$ or in $P_{p', r', q', t'}$. The result now follows from the previous proposition.

3.3. STRUCTURAL STABILITY

Given $H\epsilon R_{1, 0, 0, 1}$ it is convenient, for many purposes, to consider a slightly more restricted definition of stability.

(3.47). DEFINITION. $H(z, w) = [p(z, w)]/[q(z, w)]\epsilon R_{1, 0, 0, 1}$, where p and q are relatively prime polynomials, is said to be structurally stable if $q(z, w) \neq 0$ for $(z, w)\epsilon \bar{U}^2$ where $U = \{u\epsilon\mathbb{C}: |u| < 1\}$ and $\bar{U}$ is its closure.

It follows from (3.23) that $H\epsilon S_{p, r, q, t}$ is BIBO stable if and only if $\Phi(H)$ is BIBO stable.

(3.48). DEFINITION. $H\epsilon R_{p, r, q, t}$ is said to be *structurally stable* if $\Phi(H)$ is structurally stable.

This definition says that in any irreducible representation $[b(z, w)]/[a(z, w)]$ of $H(a, b, \epsilon P_{p, r, q, t})$, (b/a) is structurally stable if and only if it is BIBO stable and b and a have no common zeroes on T^2. (Note that by (3.35) and (3.36), $(z, w)\epsilon T^2$ if and only if $(\alpha, \beta)\epsilon T^2$). By (3.46), therefore, definition (3.48) is independent of which C_c (and therefore which Φ) we choose (for in general there will be infinitely many causality cones containing the support of $a(z, w)$ and $b(z, w)$).

(3.49). NOTATION. We let $R^s_{p, r, q, t}$ stand for the ring of structurally stable elements of $R_{p, r, q, t}$. Similarly $\bar{R}^s_{p, r, q, t}$ will stand for the structurally stable elements of $\bar{R}_{p, r, q, t}$.

We will denote by $\mathbb{R}_s[z, w]$ those polynomials which have no zeroes in $\bar{U}^2$.

3.4. MULTI-INPUT/MULTI-OUTPUT SYSTEMS

A discrete multi-input/multi-output (MIMO) 2–D system is described by means of an equation

$$\{y(m, n)\} = F(\{(x(m, n)\}) \tag{3.50}$$

where now the input $\{x(m, n)\}$ and output $\{y(m, n)\}$ are doubly indexed vector sequences.

As in the scalar case one can put conditions on the system such as linearity, shift invariance, causality and rationality.

The systems we consider will be characterized by their *transfer matrix* H which will be assumed to be in $R^{m \times k}$ where R is one of the rings $R_{p, q, r, t}$, $\bar{R}_{p, q, r, t}$, $R^s_{p, q, r, t}$ or $\bar{R}^s_{p, q, r, t}$.

Conditions for a matrix to be over any of these rings for $p = t = 1$, $q = r = 0$ ((strictly) causal and structurally stable (strictly) causal transfer matrices) will be given subsequently in terms of a matrix fraction description.

3.5. STABILIZATION OF SCALAR FEEDBACK SYSTEMS

Let

$$p(z, w) = \frac{n(z, w)}{d(z, w)}\epsilon\bar{R}_{1, 0, 0, 1} \tag{3.51}$$

where

$$n(z, w), d(z, w) \in \mathbb{R}[z, w]. \tag{3.52}$$

The assumption that p is strictly causal as opposed to just causal is made to avoid the possibility of delay free loops within the feedback system (see, for example, Cadzow [3.4]). Since any $h \in R_{1, 0, 0, 1}$ can be split into its strictly causal part and its direct coupling, this is not restrictive. We see from (3.16) that, in terms of $n(z, w)$ and $d(z, w)$, the restriction that $p \in \bar{R}_{1, 0, 0, 1}$ becomes

$$d(0, 0) \neq 0 \tag{3.53}$$

$$n(0, 0) = 0 \tag{3.54}$$

We also assume that $n(z, w)$ and $d(z, w)$ have no common factors.

Note, however, that n and d will in general have zeroes in common but by the coprimeness of n and d we have that

$$n(z, w), d(z, w)$$
have only a finite number of common zeroes. \hfill (3.55)

The feedback system we will consider is shown in Figure 3.1.

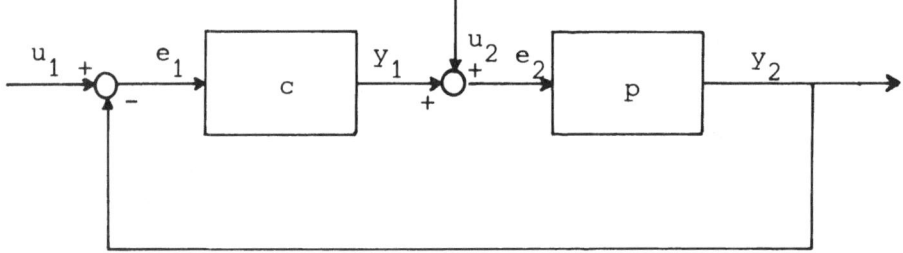

Fig. 3.1.   Scalar feedback system.

When we speak of a feedback system we will always mean one with the above structure. The approach and notation used will essentially be that of Desoer *et al.* [3.5]. We will refer to $p = p(z, w)$ as (the transfer function of) the plant and to $c = c(z, w)$ as (the transfer function of) the compensator. For the time being we assume $c \in R_{1, 0, 0, 1}$ although this restriction will be lifted in the next chapter.

Let

$$\mathbf{u} = \begin{bmatrix} u_1 \\ u_2 \end{bmatrix}, \; \mathbf{e} = \begin{bmatrix} e_1 \\ e_2 \end{bmatrix}, \; \mathbf{y} = \begin{bmatrix} y_1 \\ y_1 \end{bmatrix}.$$

Then (see [3.5])

$$\mathbf{e} = H_{eu}\mathbf{u}$$

where

$$H_{eu} = \begin{bmatrix} h_{e_1u_1} & h_{e_1u_2} \\ h_{e_2u_1} & h_{e_2u_2} \end{bmatrix} = \begin{bmatrix} \dfrac{1}{1+pc} & \dfrac{-p}{1+pc} \\ \dfrac{c}{1+pc} & \dfrac{1}{1+pc} \end{bmatrix} \tag{3.56}$$

and

$$\mathbf{y} = H_{yu}\,\mathbf{u}$$

where

$$H_{yu} = K(H_{eu} - 1) \tag{3.57}$$

and

$$K \text{ is the matrix } \begin{bmatrix} 0 & 1 \\ -1 & 0 \end{bmatrix}. \tag{3.58}$$

Consequently (Desoer and Chan [3.6]) we can restrict our attention just to H_{eu} in discussing the stability of the feedback system.

Now let us represent $c(z, w)$ as a quotient of relatively prime polynomials.

$$c(z, w) = \frac{x(z, w)}{y(z, w)} \tag{3.59}$$

Then it easily follows from (3.57) that

$$h_{e_1u_1} = \frac{yd}{yd + xn} \tag{3.60}$$

$$h_{e_2u_1} = \frac{xd}{yd + xn} \tag{3.61}$$

$$h_{e_1 u_2} = \frac{-ny}{yd + xn} \tag{3.62}$$

$$h_{e_2 u_2} = \frac{yd}{yd + xn}. \tag{3.63}$$

(3.64). DEFINITION. We will say p is *stabilizable* if we can find $c(z, w)$ such that the $h_{e_i u_j}$ $(i, j = 1, 2)$ are each in $R^s_{1,\, 0,\, 0,\, 1}$.

We wish to characterize those compensators $c(z, w) \epsilon R_{1,\, 0,\, 0,\, 1}$ which stabilize $p = (n/d)$.

Now clearly if we can find $x(z, w)$ and $y(z, w)$ such that

$$yd + xn \quad \text{is in} \quad \mathbb{R}_s[z, w]$$

then $[x(z, w)]/[y(z, w)]$ is one such compensator. Suppose $x(z, w)$ and $y(z, w)$ are such that $yd + xn$ is not a stable polynomial. Can we still achieve stability due to some cancellation in Equations (3.60) to (3.63)?

Suppose $yd + xn = r_1 r_2$ where $r_1 \epsilon \mathbb{R}_s[z, w]$ $r_2 \epsilon \mathbb{R}[z, w] \setminus \mathbb{R}_s[z, w]$. Now in Equation (3.61), the only way cancellation can occur is if x and d have a common factor since n and d are relatively prime, and x and y are relatively prime. Therefore for $h_{e_2 u_1}$ to be (structurally) stable every irreducible factor of r_2 must divide both d and x; (in practice, though, such cancellation cannot be perfectly achieved and any small fluctuation in the system will introduce instabilities). But a similar argument for the stability of $h_{e_1 u_2}$ would imply that every factor of r_2 must also divide both n and y contradicting the relative primeness of n and d, and x and y. So we have the following proposition:

(3.65). PROPOSITION. $c(z, w) = [x(z, w)]/[y(z, w)]$ will stabilize the feedback system given in Figure 3.1 if and only if

$$yd + xn \quad \text{is in} \quad \mathbb{R}_s[z, w].$$

So to determine whether a given plant $p(z, w)$ is stabilizable we must look at the set of all $yd + xn$ as y and x range over $\mathbb{R}[z, w]$. Note that for x/y to be causal we must also specify that $y(0, 0) \neq 0$; but in fact this follows automatically from our assumption that $n(0, 0) = 0$. For if $\Phi(z, w) \epsilon \mathbb{R}_s[z, w]$ is such that

$$yd + xn = \phi.$$

Then

$$y(0, 0) \, d(0, 0) = \phi(0, 0).$$

Therefore

$$y(0, 0) = \frac{\phi(0, 0)}{d(0, 0)} \neq 0.$$

So $p(z, w)$ is stabilizable by means of a causal compensator $[x(z, w)]/[y(z, w)]$ if and only if the ideal $\langle d(z, w), n(z, w) \rangle$ ($\subset \mathbb{R}[z, w]$) intersects $\mathbb{R}_s[z, w]$ non trivially. Thus we get the following theorem.

(3.66) THEOREM. $p(z, w) = [n(z, w)]/[d(z, w)]$ *is stabilizable by means of a causal compensator* $[x(z, w)]/[y(z, w)]$ *if and only if* $n(z, w)$ *and* $d(z, w)$ *have no common zeroes in* $\bar{U}^2$.

Proof. Necessity: If $n(z, w)$ and $d(z, w)$ have a common zero $(z_0, w_0) \epsilon \bar{U}^2$ i.e. $n(z_0, w_0) = d(z_0, w_0) = 0$ and $|z_0| \leq 1$, $|w_0| \leq 1$, then for any $\phi(z, w) \epsilon \mathbb{R}[z, w]$ with $\phi = yd + xn$ we have

$$\phi(z_0, w_0) = y(z_0, w_0)d(z_0, w_0) + x(z_0, w_0)n(z_0, w_0) = 0$$

and therefore ϕ cannot be in $\mathbb{R}_s[z, w]$.

SUFFICIENCY: Suppose $n(z, w)$ and $d(z, w)$ have no common zero in $\bar{U}^2$.
Let us list the common zeroes of n and d as

$$(z_1, w_1), (z_2, w_2) \ldots, (z_s, w_s), (z_{s+1}, w_{s+1}) \ldots (z_t, w_t)$$

where we can choose the indexing such that

$$|z_i| > 1 \quad i = 1, \ldots s$$
$$|w_i| > 1 \quad i = s + 1, \ldots, t.$$

Now consider the real polynomial

$$\psi(z, w) = \prod_{i=1}^{s} (z - z_i)(z - z_i^*) \prod_{i=s+1}^{t} (w - w_i)(w - w_i^*)$$

where $*$ denotes complex conjugation.
Clearly

$$\psi(z_i, w_i) = 0 \quad i = 1, \ldots, t.$$

So $\psi(z, w)$ vanishes whenever both $d(z, w)$ and $n(z, w)$ do.

Therefore by Hilbert's Nullstellensatz, there exists

$$N \epsilon \mathbb{Z}_+ \setminus \{o\} \quad \text{such that} \quad \psi^N \epsilon \langle d, n \rangle$$

where $\langle d, n \rangle$ denotes the polynomial ideal generated by d and n.
In other words, there exist $x(z, w)$, $y(z, w) \epsilon \mathbb{R}[z, w]$ such that

$$yd + xn = \psi^N.$$

Now ψ and hence ψ^N clearly has no zeroes in $\bar{U}^2$. Hence ψ^N is in $\mathbb{R}_s[z, w]$ and p is stabilizable.

3.6. CHARACTERIZATION OF STABILIZERS FOR SCALAR SYSTEMS

In the previous section we found a necessary and sufficient condition for the stabilizability of a given transfer function $p(z, w)$ and also found a compensator which would achieve this stabilization. Now we wish to characterize in some way the set of all such compensators.

We use a technique which is becoming increasingly popular in the 1–D literature (see in particular Desoer *et al.* [3.5] and Vidyasagar [3.7]); that is, instead of regarding the compensator $c(z, w)$ as a quotient of polynomials, we consider c as a quotient of elements taken from a different ring, in our case the ring $R^s_{1, 0, 0, 1}$, of (structurally) stable causal 2–D rational transfer functions. It is clear that the ring of polynomials $\mathbb{R}[z, w]$ is contained in $R^s_{1, 0, 0, 1}$. So the representation n/d of p as a quotient of polynomials is also a representation as a quotient of elements in $R^s_{1, 0, 0, 1}$.

Now in the axiomatic approach of Desoer *et al.* [3.5], the fundamental hypothesis required for their characterization theorem for compensators is the existence of a coprimeness equation of the type

$$vd + un = 1$$

where $p = nd^{-1}$ is a representation of the plant p as a quotient of elements from the ring one is working in. If we work in the ring of polynomials $\mathbb{R}[z, w]$ then such an equation will not in general hold, since even if n and d are relatively prime they will still have common zeroes which must then also appear on the right hand side of the equation. So what we do is loosen the restriction that u and v be polynomials. This leads to the following proposition.

(3.67). PROPOSITION. *Let n, d∈ℝ[z, w] be relatively prime polynomials. Then there exist u, v∈$R^s_{1, 0, 0, 1}$ such that*

$$vd + un = 1$$

if and only if n and d have no common zeroes in $\bar{U}^2$.

Proof. Suppose $vd + un = 1$ for some u, $v∈R^s_{1, 0, 0, 1}$.
Let

$$u = \frac{u_1}{u_2} \qquad v = \frac{v_1}{v_2}$$

where

$$u_1, v_1∈ℝ[z, w]$$

and

$$u_2, v_2∈ℝ_s[z, w].$$

Then

$$u_2 v_1 d + v_2 u_1 n = u_2 v_2.$$

Now let $(z_0, w_0)∈C^2$ be any common zero of n and d. Then either $u_2(z_0, w_0) = 0$ or $v_2(z_0, w_0) = 0$ which implies that $(z_0, w_0)∉\bar{U}^2$ since $u_2, v_2∈ℝ_s[z, w]$.

Conversely, suppose n and d have no common zero in $\bar{U}^2$. Then we know from the previous section that there exist x, $y∈ℝ[z, w]$ and $\phi∈ℝ_s[z, w]$ such that

$$yd + xn = \phi.$$

Therefore, taking $u = (x/\phi)$, $v = (y/\phi)$ we get $vd + un = 1$ with u, $v∈R^s_{1, 0, 0, 1}$.

Now combining Proposition (3.67) with Theorem (3.66) gives us

(3.68). PROPOSITION. *If n, d∈ℝ[z, w] are relatively prime, then n/d is stabilizable if and only if there exist u, v∈$R^s_{1, 0, 0, 1}$ such that*

$$vd + un = 1.$$

This immediately leads via the work of Desoer *et al.* [3.5] to a characterization of the stabilizing compensators. For completeness we present all of the relevant details here.

(3.69). THEOREM. *Let n, d∈ℝ[z, w] be relatively prime polynomials with no common zeroes in $\bar{U}^2$. Let u, v∈$R^s_{1, 0, 0, 1}$ be such that vd + un = 1. Then for any s(z, w)∈$R^s_{1, 0, 0, 1}$ the compensator*

$$c = \frac{-sd + u}{sn + v} \tag{3.70}$$

is causal and gives rise to a (structurally) stable feedback system. Conversely if c(z, w) is a causal (rational) compensator which gives rise to a stable feedback system, c can be written in the form of (3.70).

Proof. Let

$$s∈R^s_{1, 0, 0, 1} \text{ be arbitrary.}$$

Let

$$s = \frac{s_1}{s_2}, \qquad v = \frac{v_1}{v_2}, \qquad u = \frac{u_1}{u_2}$$

where

$$s_1, v_1, u_1∈ℝ[z, w]$$

and

$$s_2, v_2, u_2∈R_s[z, w].$$

Then

$$c = \frac{-sd + u}{sn + v}$$

$$= \frac{x(z, w)}{y(z, w)}$$

where

$$x = -s_1 v_2 u_2 d + s_2 v_2 u_1$$
$$y = s_1 v_2 u_2 n + s_2 v_1 u_2.$$

Now

$$u_2 v_1 d + v_2 u_1 n = u_2 v_2$$

and since $n(0, 0) = 0$ and $v_2(0, 0)$, $u_2(0, 0)$ are both different from 0, we have that $v_1(0, 0) \neq 0$

So

$$y(0, 0) = s_2(0, 0)\, v_1(0, 0)\, u_2(0, 0) \neq 0$$

and c is therefore a causal (though not necessarily stable) transfer function.

Furthermore,

$$yd + xn$$
$$= (s_1 v_2 u_2 n + s_2 v_1 u_2)d + (-s_1 v_2 u_2 d + s_2 v_2 u_1)n$$
$$= s_2(v_1 u_2 d + v_2 u_1 n) = s_2 u_2 v_2.$$

So, from (3.61) to (3.64) we see that the feedback system is stable.

Conversely, suppose that $c(z, w)$ is a causal (rational) transfer function which stabilizes the plant. Represent $c(z, w)$ as a quotient of relatively prime polynomials

$$c(z, w) = \frac{x(z, w)}{y(z, w)} \qquad y(0, 0) \neq 0.$$

Then, from Proposition (3.65) we have that

$$yd + xn = \phi \epsilon \mathbb{R}_s[z, w]$$

i.e.

$$\eta = \frac{y}{\phi}, \qquad \xi = \frac{x}{\phi} \tag{3.71}$$

is a solution of

$$\eta d + \xi n = 1 \quad \text{where} \quad \eta, \xi \epsilon R^s_{1, 0, 0, 1}. \tag{3.72}$$

The homogeneous equation

$$\eta^h d + \xi^h n = 0.$$

(3.73). has general solution $\eta^h = sn$, $\xi^h = -sd$ where $s\epsilon R^s_{1, 0, 0, 1}$ is arbitrary.

For certainly $\eta^h = sn$ $\xi^h = -sd$ is a solution of the homogeneous equation;

and if

$$\eta^h d + \xi^h n = 0 \quad \text{with} \quad \eta^h, \xi^h \epsilon R^s_{1, 0, 0, 1}$$

then

$$\eta^h = -\xi^h \frac{n}{d} = \frac{-\xi^h}{d} n$$

$$\xi^h = \frac{\xi^h}{d} \cdot d = -\left(\frac{-\xi^h}{d}\right) d.$$

So we just need to show $s = (-\xi^h/d)$ is in $R^s_{1,\,0,\,0,\,1}$; this follows since

$$vd + un = 1$$

and therefore

$$-\frac{\xi^h}{d} = -v\xi^h - un\frac{\xi^h}{d} = -v\xi^h + u\eta^h \epsilon R^s_{1,\,0,\,0,\,1}.$$

So from (3.73) and the hypothesis of the theorem, the general solution of

$$\eta d + \xi n = 1 \quad \text{for} \quad \eta, \, \xi \epsilon R^s_{1,\,0,\,0,\,1}$$

is

$$\eta = v + sn, \qquad \xi = u - sd \qquad s \epsilon R^s_{1,\,0,\,0,\,1}.$$

Therefore, there exists $s \epsilon R^s_{1,\,0,\,0,\,1}$ such that

$$\frac{y}{\phi} = v + sn \qquad \frac{x}{\phi} = u - sd$$

$$c = \frac{x}{y} = \frac{x}{\phi} / \frac{y}{\phi} = \frac{-sd + u}{sn + v}.$$

(3.74). *Corollary: Let n and d be relatively prime polynomials with no common zeroes in $\bar{U}^2$. Let x, $y \epsilon \mathbb{R}[z, w]$ be such that*

$$yd + xn \epsilon R_s[z, w].$$

Then the stabilizing compensators are characterized by the expression

$$c = \frac{-s_1 d + s_2 x}{s_1 n + s_2 y} \tag{3.75}$$

where

$$s_1 \epsilon \mathbb{R}[z, w]$$

and

$s_2 \in \mathbb{R}_s[z, w]$ are arbitrary.

Proof. If $yd + xn = \phi$, Let $v = y/\phi$, $u = x/\phi$ and use the previous theorem.

To summarize: given a plant $p(z, w) = [n(z, w)]/[d(z, w)]$ where d and n are relatively prime, proceed as follows:

(1) If n and d have common zeroes in $\bar{U}^2$ then p cannot be stabilized by a causal (rational) compensator.

(2) If n and d have no common zeroes in $\bar{U}^2$, then find $x(z, w)$ and $y(z, w)$ such that $yd + xn$ is stable. A method for doing this is discussed in the proof of (3.66) using the constructive approach in Chapter 6 to determine whether a given polynomial is in a given polynomial ideal specified by a finite number of generators.

(3) The compensators are then given by

$$c = \frac{-s_1 d + s_2 x}{s_1 n + s_2 y}$$

where s_1 is an arbitrary element of $\mathbb{R}_s[z, w]$.

(4) From (3.56) we then obtain with the above c

$$H_{eu} = \begin{bmatrix} \dfrac{d(s_1 n + s_2 y)}{s_2(yd + xn)} & \dfrac{-n(s_1 n + s_2 y)}{s_2(yd + xn)} \\[3mm] \dfrac{d(-s_1 d + s_2 x)}{s_2(yd + xn)} & \dfrac{d(s_1 n + s_2 y)}{s_2(yd + xn)} \end{bmatrix}.$$

(3.76). *Example*: (a)

$$\frac{n(z, w)}{d(z, w)} = \frac{zw - z - 2w}{z^2 w^2 - 2zw^2 - 2zw + 4z + 4}.$$

We construct a Gröbner basis (see Chapter 6 of this text) for the polynomials $z^2 w^2 - 2zw^2 - 2zw + 4z + 4$ and $zw - z - 2w$ and in the process generate the polynomials $z^2 + 4z + 4 \in \mathbb{R}_s[z, w]$. In fact we have:

$$z^2 + 4z + 4 = d(z, w) - z(w + 1)n(z, w).$$

The compensators are then characterized by the expression

$$\frac{-s_1(z, w)(z^2w^2 - 2zw^2 - 2zw + 4z + 4) - s_2(z, w)z(w + 1)}{s_1(z, w)(zw - z - 2w) + s_2(z, w)}$$

where

$$s_1 \epsilon \mathbb{R}[z, w] \quad \text{and} \quad s_2 \epsilon \mathbb{R}_s[z, w]$$

(b)

$$\frac{n(z, w)}{d(z, w)} = \frac{z + w}{1 - z + w}.$$

$n(z, w)$ and $d(z, w)$ have a common zero at the point $(\frac{1}{2}, -\frac{1}{2}) \epsilon \bar{U}^2$. Consequently we cannot stabilize this transfer function by means of a causal compensator; however we will see in Section 3.9 (example (3.143)) that it can be stabilized by means of a weakly causal compensator.

3.7. STABILIZATION OF STRICTLY CAUSAL TRANSFER MATRICES

(3.77) Let P be an $m \times k$ matrix with entries

$$P_{ij}(z, w) = \frac{n_{ij}(z, w)}{d_{ij}(z, w)}$$

where

$$n_{ij}, d_{ij} \epsilon \mathbb{R}[z, w] \tag{3.78}$$

and are relatively prime

$$i = 1, \ldots m, \quad j = 1, \ldots k.$$

We will be interested in those P for which

$$d_{ij}(0, 0) \neq 0 \tag{3.79}$$

and

$$n_{ij}(0, 0) = 0 \tag{3.80}$$

$$i = 1, \ldots, m, \quad j = 1, \ldots k.$$

Recall (Section 3.4) that such a P is called a strictly causal rational transfer matrix. The following two propositions illustrate the relationship of Conditions (3.79) and (3.80) with matrix fraction descriptions of P.

(3.81). PROPOSITION. (i) *If P satisfies (3.79)* then any irreducible left matrix fraction description (LMFD) $A^{-1}B$ of P has (det A) (0, 0) $\neq$ 0. (see references [3.8], [3.10] for definitions of LMFD, irreducible LMFD or coprime LMFD)

(ii) *Conversely, if one can find a (not necessarily irreducible) LMFD $A^{-1}B$ of P such that (det A) (0, 0) $\neq$ 0, then P satisfies (3.79).*

Proof. Let

$$d_i = \text{LCM}\{d_{ij}: j = 1, \dots k\} \qquad i = 1, \dots, m$$

$$D_1 = \text{diag}\{d_1, \dots, d_m\}$$

and

$$N_1 = \left[n_{ij} \frac{d_i}{d_{ij}} \right].$$

Clearly, $P = D_1^{-1}N_1$

Form $[D_1\ N_1]$ and extract the greatest common left divisor L of $[D_1\ N_1]$ using the procedure in [3.8] to obtain

$$[D_1\ N_1] = L[D_L\ N_L].$$

(i) It is clear that (det D_1)(0, 0) $\neq$ 0 (for P satisfying (3.79)) and therefore (det D_L)(0, 0) $\neq$ 0.

Given any other irreducible LMFD $A^{-1}B$ of P we know (follows transitively from results in [3.10, Theorem 5.2])

$$\det A = k \det D_L$$

where k is a non-zero constant.

(ii) $A^{-1}B = (A^+B)/(\det A)$ where A^+ is the adjoint matrix of A. The result now follows from (3.78).

(3.82). PROPOSITION. *Let $A^{-1}B$ be any (not necessarily irreducible) LMFD of P such that (det A)(0, 0) $\neq$ 0. Then the following are equivalent.*

(i) *(3.80) is satisfied.*

(ii) $b_{ij}(0, 0) = 0$ *for any entry b_{ij} of the matrix B.*

(iii) *All m^{th} order minors of [A, B], except for det A, are zero when evaluated at (0, 0).*

Proof. Follows immediately from $A[I_m\ P] = [A\ B]$.

Similar statements for the above two propositions follow of course for any right matrix fraction description (RMFD) of P.

(3.83). Let P be an $m \times k$ strictly causal rational transfer matrix. We can represent P by means of an irreducible LMFD

$$D_L^{-1}N_L \tag{3.84}$$

where from (3.81) and (3.82)

$$(\det D_L)(0, 0) \neq 0 \tag{3.85}$$

and

$$N_L(0, 0) = 0_{m \times k}. \tag{3.86}$$

Since $D_L^{-1}N_L$ is irreducible we have that the mth order minors of $[D_L N_L]$ have no common factor and hence have only a finite number of common zeroes.

3.7.1. *MIMO Feedback Systems and Their Stabilization*

As in the scalar case we will be considering a feedback system of the type shown in Figure 3.2.

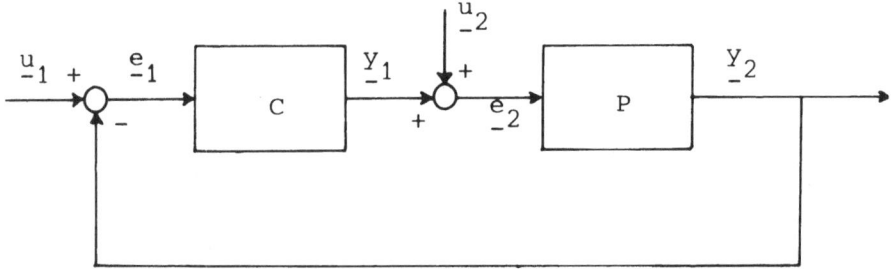

Fig. 3.2. Matrix feedback system.

Here,

$$\mathbf{y}_1, \mathbf{u}_2 \text{ and } \mathbf{e}_2 \text{ are } k \times 1 \text{ vectors}$$

$$\mathbf{y}_2, \mathbf{u}_1 \text{ and } \mathbf{e}_1 \text{ are } m \times 1 \text{ vectors}$$

and C is assumed to be a $k \times m$ causal rational transfer matrix.

Let

$$\mathbf{u} = \begin{bmatrix} \mathbf{u}_1 \\ \mathbf{u}_2 \end{bmatrix}, \; \mathbf{e} = \begin{bmatrix} \mathbf{e}_1 \\ \mathbf{e}_2 \end{bmatrix}, \; \mathbf{y} = \begin{bmatrix} \mathbf{y}_1 \\ \mathbf{y}_2 \end{bmatrix}$$

then

$$\mathbf{e} = H_{eu}\mathbf{u}$$

where

$$H_{eu} = \begin{bmatrix} h_{e_1 u_1} & h_{e_1 u_2} \\ h_{e_2 u_1} & h_{e_2 u_2} \end{bmatrix} = \begin{bmatrix} (I_m + PC)^{-1} & -(I_m + PC)^{-1}P \\ C(I_m + PC)^{-1} & I_k - C(I_m + PC)^{-1}P \end{bmatrix}. \tag{3.87}$$

Note that H_{eu} is well defined since at $(0, 0)$ P and C are both well defined (being causal) and $P(0, 0) = O_{m \times k}$. So $(I_m + PC)(0, 0) = I_m$.

We also have

$$\mathbf{y} = H_{yu}\,\mathbf{u}$$

where

$$H_{yu} = K(H_{eu} - I_{m+k}) \tag{3.88}$$

and K is the matrix

$$\begin{bmatrix} O_{k \times m} & I_k \\ -I_m & O_{m \times k} \end{bmatrix}. \tag{3.89}$$

So as in the scalar case we can restrict our attention just to H_{eu} in discussing the stability of the feedback system.

Now represent C by an irreducible RMFD

$$C = X_R\,Y_R^{-1} \tag{3.90}$$

Since C is assumed to be causal, we have from Proposition (3.81) that

$$(\det Y_R)(0, 0) \neq 0 \tag{3.91}$$

It now follows from (3.87) that

$$h_{e_1 u_1} = Y_R(D_L Y_R + N_L X_R)^{-1}D_L \tag{3.92}$$

$$h_{e_2 u_1} = X_R(D_L Y_R + N_L X_R)^{-1}D_L \tag{3.93}$$

$$h_{e_1 u_2} = -Y_R(D_L Y_R + N_L X_R)^{-1}N_L \tag{3.94}$$

$$h_{e_2 u_2} = I_k - X_R(D_L Y_R + N_L X_R)^{-1}N_L \tag{3.95}$$

We could also work in terms of an irreducible RMFD for P, and an irreducible LMFD for C;

$$P = N_R D_R^{-1} \qquad C = Y_L^{-1} X_L. \tag{3.96}$$

It is not difficult to show that

$$H_{eu} = \begin{bmatrix} I_m - P(I_k + CP)^{-1}C & -P(I_k + CP)^{-1} \\ (I_k + CP)^{-1}C & (I_k + CP)^{-1} \end{bmatrix} \tag{3.97}$$

It then follows from (3.96) that

$$h_{e_1u_1} = I_m - N_R(Y_L D_R + X_L N_R)^{-1} X_L \tag{3.98}$$

$$h_{e_2u_1} = D_R(Y_L D_R + X_L N_R)^{-1} X_L \tag{3.99}$$

$$h_{e_1u_2} = -N_R(Y_L D_R + X_L N_R)^{-1} Y_L \tag{3.100}$$

$$h_{e_2u_2} = D_R(Y_L D_R + X_L N_R)^{-1} Y_L. \tag{3.101}$$

(3.102). PROPOSITION. $\det(Y_L D_R + X_L N_R) = a \det(D_L Y_R + N_L X_R)$ *where a is a nonzero constant.*

Proof.

$$\begin{bmatrix} Y_L & X_L \\ N_L & -D_L \end{bmatrix} \begin{bmatrix} D_R & 0 \\ N_R & I_m \end{bmatrix} = \begin{bmatrix} Y_L D_R + X_L N_R & X_L \\ 0 & -D_L \end{bmatrix}$$

Therefore, since from [3.10, Theorem 5.2] $\det D_R$ = const. $\det D_L$ We have that

$$\det \begin{bmatrix} Y_L & X_L \\ N_L & -D_L \end{bmatrix} = \text{const. } \det(Y_L D_R + X_L N_R). \tag{3.103}$$

Similarly,

$$\begin{bmatrix} Y_L & X_L \\ N_L & -D_L \end{bmatrix} \begin{bmatrix} I_k & X_R \\ 0 & -Y_R \end{bmatrix} = \begin{bmatrix} Y_L & 0 \\ N_L & D_L Y_R + N_L X_R \end{bmatrix}$$

implies (since $\det Y_L$ = const. $\det Y_R$) that

$$\det \begin{bmatrix} Y_L & X_L \\ N_L & -D_L \end{bmatrix} = \text{const. } \det(D_L Y_R + N_L X_R). \tag{3.104}$$

The result now follows from (3.103) and (3.104).

(3.105). COROLLARY. *If* $P = \tilde{D}_L^{-1}\tilde{N}_L$ *and* $C = \tilde{X}_R\tilde{Y}_R^{-1}$ *are any other irreducible* M.F.D's *of* P *and* C *then* $\det(\tilde{D}_L\tilde{Y}_R + \tilde{N}_L\tilde{X}_R) = \det(D_LY_R + N_LX_R)$

Proof. Follows transitively from Proposition (3.102).

We now show that $\det(D_LY_R + N_LX_R)$ (which by the above is independent of the representations of P and C provided they are irreducible) contains all the information about the (structural) stability of the matrix H_{eu}.

(3.106). PROPOSITION. *The feedback system in Figure 3.2 is (structurally) stable if and only if* $\det(D_LY_R + N_LX_R)$ *is a stable polynomial.*

Proof. The feedback system is stable if and only if the matrix H_{eu} is stable. This is true if and only if

$$\begin{bmatrix} h_{e_1u_1} & -h_{e_1u_2} \\ h_{e_2u_1} & I_k - h_{e_2u_2} \end{bmatrix} \text{ is stable.}$$

But

$$\begin{bmatrix} h_{e_1u_1} & -h_{e_1u_2} \\ h_{e_2u_1} & I_k - h_{e_2u_2} \end{bmatrix} =$$

$$\begin{bmatrix} Y_R(D_LY_R + N_LX_R)^{-1}D_L & Y_R(D_LY_R + N_LX_R)^{-1}N_L \\ X_R(D_LY_R + N_LX_R)^{-1}D_L & X_R(D_LY_R + N_LX_R)^{-1}N_L \end{bmatrix}$$

$$= \begin{bmatrix} Y_R \\ X_R \end{bmatrix}(D_LY_R + N_LX_R)^{-1}[D_L \quad N_L]. \tag{3.107}$$

Now

since Y_R and X_R are right coprime

and D_L and N_L are left coprime,

we have that

$$\begin{bmatrix} Y_R \\ X_R \end{bmatrix} \quad \text{and} \quad D_L Y_R + N_L X_R \text{ are right coprime} \qquad (3.108)$$

and

$$[D_L \ N_L] \quad \text{and} \quad D_L Y_R + N_L X_R \text{ are left coprime} \qquad (3.109)$$

so if

$$\begin{bmatrix} Y_R \\ X_R \end{bmatrix} (D_L Y_R + N_L X_R)^{-1}[D_L \ N_L] = A^{-1}B$$

where $A^{-1}B$ is irreducible then from (3.108) and (3.109) it follows that $\det(D_L Y_R + N_L X_R) = \text{const. det } A$. The result now follows since $A^{-1}B$ is (structurally) stable if and only if det A is a stable polynomial on using (3.110) below

(3.110). THEOREM. Let $P = A_L^{-1}B_L = VT^{-1}U$, where $A_L^{-1}B_L$, VT^{-1} and $T^{-1}U$ are all irreducible. Then det $T = k \cdot \det A_L$ where k is a non-zero constant. See [3.10] for a proof.

(3.111). NOTE. (3.102) and (3.105) also follow as corollaries to (the proof of) (3.106) by uniqueness of 'denominators'.

In (3.91) we have specified that $(\det Y_R)(0, 0) \neq 0$, but if $\det(D_L Y_R + N_L X_R)$ is a stable polynomial this is not necessary:

(3.112). LEMMA. *Let N_L, D_L satisfy* (3.85) *and* (3.86).
Then if det $(D_L Y_R + N_L X_R)$ *is stable,* (det Y_R)(0, 0) $\neq 0$.
Proof. Since det $(D_L Y_R + N_L X_R)$ is stable, (det $(D_L Y_R + N_L X_R)$) $(0, 0) \neq 0$.

The result now follows from the Cauchy–Binet theorem and (3.85) and (3.86).

We now state a slightly extended version of a result due to Youla and Gnavi [3.9]. The following Proposition though also stated and proved in Chapter 7 (see Lemma 7.10) in a manner different from that in [3.9], is quoted here to facilitate comprehension of the proof of Theorem (3.114).

(3.113). PROPOSITION. Let A be an $m \times \ell$ matrix with entries in D, $m \le \ell$ (here D is an Euclidean domain)

$$\text{Let } a_{i_1 \ldots i_m} = A \begin{pmatrix} 1 & \ldots & m \\ i_1 & \ldots & i_m \end{pmatrix}^\dagger$$

and let I be the ideal of D generated by the $a_{i_1 \ldots i_m}$'s.

Then, given any $a\epsilon I$, there exists an $\ell \times m$ matrix B with entries in D such that

$$AB = aI_m.$$

(3.114). THEOREM. $P(z, w) = D_L^{-1}(z, w)N_L(z, w)$ is stabilizable by means of causal compensator if and only if the mth order minors of $[D_L \ N_L]$ have no common zero in $\bar{U}^2$.

Proof. Necessity: If the mth order minors of $[D_L \ N_L]$ have a common (z_0, w_0), then for any Y_R, X_R we have, by the Cauchy–Binet theorem, (z_0, w_0) is a zero of

$$\det(D_L Y_R + N_L X_R) \left(= \det[D_L N_L] \begin{bmatrix} Y_R \\ X_R \end{bmatrix} \right).$$

So if $|z_0| \le 1$, $|w_0| \le 1$ $\det(D_L Y_R + N_L X_R)$ cannot be stable. Sufficiency: Suppose the mth order minors of $[D_L \ N_L]$ have no common zero in $\bar{U}^2$.

Since $D_L^{-1}N_L$ is irreducible, the mth order minors of $[D_L \ N_L]$ have only a finite number of common zeroes. We can therefore, precisely as in the proof of Theorem (3.66) find a stable polynomial $\psi(z, w)$ which vanishes at each common zero of the mth order minors of $[D_L \ N_L]$. Therefore by Hilberts Nullstellensatz and by Proposition (3.113) there exist $N\epsilon\mathbb{Z}_+ \setminus \{0\}$, and Y and X, $m \times m$ and $k \times m$ matrices respectively with entries in $\mathbb{R}[z, w]$, such that $[D_L \ N_L][{}^Y_X] = \psi^N I$.

Consequently $X Y^{-1}$ is a compensator which stabilizes P, and is causal by Lemma (3.112).

3.8. CHARACTERIZATION OF STABILIZERS FOR MIMO SYSTEMS

(3.115). As in the scalar case we can characterize in a systematic way those (causal) compensators which stabilize $P = D_L^{-1}N_L$. We again

---

† This denotes the determinant of submatrix of a matrix A obtained by taking rows 1, 2, . . ., m and columns $i_1, i_2, \ldots i_m$.

assume for the remainder of this chapter that $D_L^{-1}N_L$ satisfies (3.84), (3.85), and (3.86).

(3.116). PROPOSITION. *There exist* $V \epsilon (R_{1, 0, 0, 1}^s)^{m \times m}$ *and* $U \epsilon (R_{1, 0, 0, 1}^s)^{k \times m}$ *such that*

$$D_L V + N_L U = I_m$$

if and only if the mth *order minors of* $[D_L \ N_L]$ *have no common zeroes in* $\bar{U}^2$.

  Proof. Suppose $D_L V + N_L U = I_m$, with $V \epsilon (R_{1, 0, 0, 1}^s)^{m \times m}$ and $U \epsilon (R_{1, 0, 0, 1}^s)^{k \times m}$.

  Then

$$V = \frac{V_1}{v_2} \qquad U = \frac{U_1}{u_2}$$

where $V_1 \epsilon R^{m \times m}[z, w]$, $U_1 \epsilon R^{k \times m}[z, w]$, and $v_2, u_2 \epsilon R_s[z, w]$.

  Thus $u_2 D_L V_1 + v_2 N_L U_1 = u_2 v_2 I$, so by the Cauchy–Binet theorem, any common zero $(z_0, w_0) \epsilon \mathbb{C}^2$ of the mth order minors of $[D_L \ N_L]$ must also be a zero of u_2 or v_2. Since u_2 and v_2 are in $R_s[z, w]$ such a zero cannot be in $\bar{U}^2$.

  Conversely, if the mth order minors of $[D_L \ N_L]$ have no common zeroes in $\bar{U}^2$, we know from the sufficiency part of the proof of Theorem (3.114) that there exist $Y \epsilon R^{m \times m}[z, w]$ and $X \epsilon R^{k \times m}[z, w]$ and $\phi \epsilon R_s^{m \times m}[z, w]$ such that $D_L Y + N_L X = \phi I_m$.

  Therefore taking

$$V = \frac{Y}{\phi} \qquad U = \frac{X}{\phi}$$

gives

$$D_L V + N_L U = I_m$$

with

$$V \epsilon (R_{1, 0, 0, 1}^s)^{m \times m}, \ U \epsilon (R_{1, 0, 0, 1}^s)^{k \times m}.$$

Combining proposition (3.116) with Theorem (3.114) gives.

(3.117). PROPOSITION. $D_L^{-1}N_L$ *is stabilizable by means of a causal compensator if and only if there exist* $V_R \epsilon (R_{1, 0, 0, 1}^s)^{m \times m}$, $U_R \epsilon (R_{1, 0, 0, 1}^s)^{k \times m}$ *such that* $D_L V_R + N_L U_R = I_m$.

As in the scalar case this leads, using the work of Desoer *et al.* [3.5] to a characterization of the stabilizing compensators. Again, we give all details.

(3.118). THEOREM. *Let $D_L^{-1}N_L$ and $N_R D_R^{-1}$ be irreducible representations of the strictly causal plant $P(z, w)$ such that the mth order minors of $[D_L N_L]$ have no common zeroes in $\bar{U}^2$. Let $U_R \epsilon (R_{1, 0, 0, 1}^s)^{k \times m}$ and $V_R \epsilon (R_{1, 0, 0, 1}^s)^{m \times m}$ be such that $D_L V_R + N_L U_R = I_m$ (such U_R and V_R exist by previous proposition). Then for any $S \epsilon (R_{1, 0, 0, 1}^s)^{k \times m}$ the compensator*

$$C = (-D_R S + U_R)(N_R S + V_R)^{-1} \qquad (3.119)$$

is causal and stabilizes the feedback system. Conversely, if $C(z, w)$ is a rational causal compensator which stabilizes the feedback system, C can be written in the form of (3.119).

Proof. Let $S \epsilon (R_{1, 0, 0, 1}^s)^{k \times m}$ be arbitrary. Let

$$S = \frac{S_1}{s_2}, \; V_R = \frac{V_1}{v_2}, \; U_R = \frac{U_1}{u_2}$$

where

$$S_1, \; U_1 \epsilon \mathbb{R}^{k \times m}[z, w], \; V_1 \epsilon \mathbb{R}^{m \times m}[z, w]$$
$$s_2, \; v_2, \; u_2 \epsilon \mathbb{R}_s[z, w].$$

Then

$$C = (-D_R S + U_R)(N_R S + V_R)^{-1}$$
$$= X Y^{-1}$$

where

$$X = -v_2 u_2 D_R S_1 + s_2 v_2 U_1$$
$$Y = v_2 u_2 N_R S_1 + s_2 u_2 V_1.$$

Now

$$[D_L \; N_L] \begin{bmatrix} u_2 V_1 \\ v_2 U_1 \end{bmatrix} = u_2 v_2 I_m. \qquad (3.120)$$

Since all the mth order minors of $[D_L \; N_L]$ except for det D_L are zero when evaluated at $(0, 0)$ (by (3.85) and (3.86)), and since $u_2(0, 0) \neq 0$, $v_2(0, 0) \neq 0$ (u_2 and v_2 being in $\mathbb{R}_s[z, w]$), we have from (3.120) and the Cauchy–Binet theorem that (det $V_1)(0, 0) \neq 0$.

Therefore,

$$(\det Y)(0, 0)$$
$$= \det (Y(0, 0))$$
$$= \det((v_2 u_1 N_R S_1)(0, 0) + (s_2 u_2 V_1)(0, 0))$$
$$= \det(s_2 u_2 V_1)(0, 0) \ (\text{since} N_R(0, 0) = 0$$
$$\text{by Proposition (3.82)(ii)})$$
$$= s_2^m(0, 0) \ u_2^m(0, 0)(\det V_1)(0, 0) \neq 0$$

and hence C is causal.
Furthermore

$$D_L Y + N_L X$$
$$= v_2 u_2 D_L N_R S_1 + s_2 u_2 D_L V_1 - v_2 u_2 N_L D_R S_1 + s_2 v_2 N_L U_1$$
$$= s_2 u_2 D_L V_1 + s_2 v_2 N_L U_1 \quad (\text{since } D_L N_R = N_L D_R)$$
$$= s_2 u_2 v_2 I_m \quad (\text{see (3.120)})$$

So, from (3.92), (3.93), (3.94), and (3.95) we see that the feedback system is stable.

Conversely, suppose $C(z, w)$ is a rational causal transfer matrix giving rise to a stable feedback system.

Let XY^{-1} be a irreducible RMFD of C.

Since we are assuming C causal we have from Proposition (3.81) that

$$(\det Y)(0, 0) \neq 0. \tag{3.121}$$

From Proposition (3.106) we have that $(D_L Y + N_L X) = \Phi$ where $\det \Phi \epsilon \mathbb{R}_s[z, w]$.

(3.122). Therefore $H = Y\Phi^{-1} \quad \Xi = X\Phi^{-1}$ is a solution of

$$D_L H + N_L \Xi = I_m \quad \text{where } H\epsilon(R_{1, 0, 0, 1}^s)^{m \times m}$$
$$\text{and} \quad \Xi\epsilon(R_{1, 0, 0, 1}^s)^{k \times m} \tag{3.123}$$

The homogeneous equation

$$D_L H^h + N_L \Xi^h = 0_{m \times m}$$

has general solution $H^h = N_R S$

$$\Xi^h = -D_R S \tag{3.124}$$

where $S\epsilon(R^s_{1, 0, 0, 1})^{k \times m}$ is arbitrary.

For certainly (3.124) is a solution (since $N_L D_R = D_L N_R$) for arbitrary $S\epsilon(R^s_{1, 0, 0, 1})^{k \times m}$; and if $D_L H^h + N_L \Xi^h = 0_{m \times m}$ where

$$H^h \epsilon(R^s_{1, 0, 0, 1})^{m \times m}, \qquad \Xi^h \epsilon(R^s_{1, 0, 0, 1})^{k \times m}$$

then

$$H^h = -D_L^{-1} N_L \Xi^h = N_R(-D_R^{-1} \Xi^h)$$
$$\Xi^h = -D_R(-D_R^{-1} \Xi^h).$$

So we just need to show

$$-D_R^{-1} \Xi^h \epsilon(R^s_{1, 0, 0, 1})^{k \times m}; \qquad \text{this follows since}$$

$$V_L D_R + U_L N_R = I_k \quad \text{for some}$$
$$U_L \epsilon(R^s_{1, 0, 0, 1})^{k \times m}, V_L \epsilon(R^s_{1, 0, 0, 1})^{k \times k}$$

and therefore

$$-D_R^{-1} \Xi^h = -(V_L D_R + U_L N_R) D_R^{-1} \Xi^h$$
$$= -V_L \Xi^h - U_L N_R D_R^{-1} \Xi^h$$
$$= -V_L \Xi^h + U_L H^h \epsilon(R^s_{1, 0, 0, 1})^{k \times m}.$$

So from (3.124), the general solution of $D_L H + N_L \Xi = I_m$ for $H\epsilon(R^s_{1, 0, 0, 1})^{m \times m}, \Xi\epsilon(R^s_{1, 0, 0, 1})^{k \times m}$ is

$$H = V_R + N_R S \qquad \Xi = U_R - D_R S \qquad (3.125)$$

where

$$S\epsilon(R^s_{1, 0, 0, 1})^{k \times m}.$$

Therefore, from (3.122) and (3.123) $\exists S\epsilon(R^s_{1, 0, 0, 1})^{k \times m}$ such that

$$Y\Phi^{-1} = V_R + N_R S$$
$$X\Phi^{-1} = U_R - D_R S.$$

Therefore

$$C = XY^{-1} = X\Phi^{-1} \Phi Y^{-1}$$
$$= (X\Phi^{-1})(Y\Phi^{-1})^{-1}$$

(3.126). COROLLARY. Let N_L, D_L, N_R, D_R be as in the previous theorem. Let $X_R \epsilon \mathbb{R}^{k \times m}[z, w]$ and $Y_R \epsilon \mathbb{R}^{m \times m}[z, w]$ be such that

$$\det(D_L Y_R + N_L X_R) \epsilon \mathbb{R}_S[z, w].$$

Then the stabilizing compensators are characterized by

$$C = (-D_R R_1 + r_2 X_R) (D_R R_1 + r_2 Y_R)^{-1}$$

where

$$R_1 \epsilon \mathbb{R}^{k \times m}[z, w]$$

and

$$r_2 \epsilon \mathbb{R}_S[z, w] \text{ are arbitrary.}$$

3.9. STABILIZATION OF WEAKLY CAUSAL SYSTEMS

(3.127). NOTATION. Let

$$p, q, r, t \epsilon \mathbb{Z}_+$$

such that

$$pt - qr = 1.$$

We denote the set $\{(z, w) \epsilon \mathbb{C}^2 : |z|^t \le |w|^q, |w|^p \le |z|^r\}$ by

$$V(p, r, q, t).$$

(3.128). PROPOSITION. Let

$$\frac{b(z, w)}{a(z, w)}$$

be an irreducible representation (see (3.42)) of $p(z, w) \epsilon R_{p, r, q, t}$ where a, $b \epsilon P_{p, r, q, t}$ and $a(z, w)$ has a non-zero constant term. Then $p \epsilon R^s_{p, r, q, t}$ if and only if $a(z, w)$ has no zeroes in $V(p, r, q, t)$.

(3.129). By a zero of $a(z, w)$ we mean a pair $(z_0, w_0) \epsilon \mathbb{C}^2$ for which each term in $a(z, w)$ is well defined and such that $a(z_0, w_0) = 0$. Note that by this definition $(0, 0)$ cannot be a zero of $a(z, w)$.

Proof of Proposition.

$$p(z, w) = \frac{b(z, w)}{a(z, w)}.$$

We can write (see (3.41)) $b(z, w)$ and $a(z, w)$ as polynomials in $z^t w^{-q}$ and $z^{-r} w^p$:

$$a(z, w) = a_0(z^t w^{-q}, z^{-r} w^p) \tag{3.130}$$

$$b(z, w) = b_0(z^t w^{-q}, z^{-r} w^p) \tag{3.131}$$

and

$$\Phi\left(\frac{b(z, w)}{a(z, w)}\right) = \frac{\Phi(b(z, w))}{\Phi(a(z, w))} = \frac{\Phi(b_0(z^t w^{-q}, z^{-r} w^p))}{\Phi(a_0(z^t w^{-q}, z^{-r} w^p))} = \frac{b_0(\alpha, \beta)}{a_0(\alpha, \beta)}$$

$$\tag{3.132}$$

Now if there exists $(z_0, w_0) \epsilon V(p, r, q, t)$ such that $a(z_0, w_0) = 0$, then setting

$$\alpha_0 = z_0^t w_0^{-q}, \; \beta_0 = z_0^{-r} w_0^p$$

we get that

$$(\alpha_0, \beta_0) \epsilon \bar{U}^2 \quad \text{and} \quad a_0(\alpha_0, \beta_0) = 0.$$

Therefore, from the definition of structural stability in $R_{p, r, q, t}$ we have

$$\frac{b(z, w)}{a(z, w)} \notin R^s_{p, r, q, t}.$$

Conversely, suppose

$$\frac{b(z, w)}{a(z, w)} \notin R^s_{p, r, q, t}.$$

Then

$$a_0(\alpha_0, \beta_0) = 0 \quad \text{for some} \quad (\alpha_0, \beta_0) \epsilon \bar{U}^2.$$

By the continuity of the zeroes of a polynomial we can assume that

$$\alpha_0 \neq 0 \quad \text{and} \; \beta_0 \neq 0.$$

Then, defining

$$z_0 = \alpha_0^p \beta_0^q, \qquad w_0 = \alpha_0^r \beta_0^t$$

we obtain a zero (z_0, w_0) of $a(z, w)$ lying in $V(p, r, q, t)$.

We now consider the problem of stabilizing a plant $p(z, w)$ in $\bar{R}_{p, r, q, t}$ using a compensator $c(z, w)$ in $R_{p, r, q, t}$. We use the same feedback configuration as in Figure 3.1. Because of the ring isomorphism Φ which maps $\bar{R}_{p, r, q, t}$ bijectively onto $\bar{R}_{1, 0, 0, 1}$ and $R_{p, r, q, t}$ bijectively onto $R_{1, 0, 0, 1}$, we can convert the problem to the causal case, solve it using the methods of Section 3.5 and then convert back again.

So from Theorem (3.66) we have the following proposition.

(3.133). PROPOSITION. The plant $p \epsilon \bar{R}_{p, r, q, t}$ is stabilizable by means of a compensator $c \epsilon R_{p, r, q, t}$ if and only if

$$\Phi(a)(\alpha, \beta) \quad \text{and} \quad \Phi(b)(\alpha, \beta)$$

have no common zeroes in $\bar{U}^2$ (where $[b(z, w)]/[a(z, w)]$ is irreducible).

One might expect that in terms of $a(z, w)$ and $b(z, w)$ the condition for stabilizability is that $a(z, w)$ and $b(z, w)$ have no common zeroes in $V(p, r, q, t)$. Unfortunately this is not the case since $\Phi(a)(\alpha, \beta)$ and $\Phi(b)^t(\alpha, \beta)$ may have zeroes in $\bar{U}^2$ of the form $(\alpha_0, 0)$ or $(0, \beta_0)$ and these are not describable in terms of z and w.

(3.134). EXAMPLE. $p = t = q = 1 \qquad r = 0$

$$\frac{b(z, w)}{a(z, w)} = \frac{z^2 w^{-2} + z w^{-1} + w}{1 + z w^{-1} + w} \; \epsilon \bar{R}_{1, 0, 1, 1}.$$

If it is easily shown that $b(z, w)$ and $a(z, w)$ only have a common zero at $(-2, -2)$ which does not lie in $V(1, 0, 1, 1)$.
However

$$\Phi(b)(\alpha, \beta) = \alpha^2 + \alpha + \beta$$

and

$$\Phi(a)(\alpha, \beta) = 1 + \alpha + \beta$$

have common zeroes at $(1, -2)$ and $(-1, 0)$. Therefore $[b(z, w)]/[a(z, w)]$ is not stabilizable in $R_{1, 0, 1, 1}$.

If, given p, r, q, t with $pt - qr = 1$, we let $a_{p, r}(z, w)$ denote that portion of $a(z, w)$ involving just powers of $w^p z^{-r}$ (including the constant term) and $a_{q,\,t}(z, w)$ denote that portion of $a(z, w)$ involving just powers of $z^t w^{-q}$ (including the constant term), and similarly define $b_{p,\,r}(z, w)$ and $b_{q,\,t}(z, w)$, then the criterion for stabilizability clearly becomes

(3.135). (i) $b(z, w)$ and $a(z, w)$ have no common zero in $V(p, r, q, t)$

(ii) $b_{q,\,t}(z, w)$ and $a_{q,\,t}(z, w)$ have no common zero in

$$\frac{|z|^t}{|w|^q} \le 1$$

(iii) $b_{p,\,r}(z, w)$ and $a_{p,\,r}(z, w)$ have no common zero in

$$\frac{|w|^p}{|z|^r} \le 1.$$

Given a weakly causal rational transfer function the region of support of the numerator and denominator is not contained in an uniquely defined causality cone. In this section we consider the following problem: given a plant $p(z, w) \epsilon \bar{R}_{p,\,r,\,q,\,t}$ which is not stabilizable in $R_{p,\,r,\,q,\,t}$, find

$$p^*, r^*, q^*, t^*$$

such that

$$H_{p,\,r} \cap H_{q,\,t} \subset H_{p^*,\,r^*} \cap H_{q^*,\,t^*}$$

(and therefore $p(z, w) \epsilon \bar{R}_{p^*,\,r^*,\,q^*,\,t^*}$) and $p(z, w)$ is stabilizable in $R_{p^*,\,r^*,\,q^*,\,t^*}$.

(3.136). EXAMPLE. Consider the $p(z, w)$ given in Example (3.134).

i.e. $$\frac{b(z, w)}{a(z, w)} = \frac{z^2 w^{-2} + zw^{-1} + w}{1 + zw^{-1} + w}$$

We chose as our region of support $H_{1,\,0} \cap H_{1,\,1}$. However, we found that while conditions (i) and (iii) of (3.135) were satisfied, condition (ii) was not since $z^2 w^{-2} + zw^{-1}$ and $1 + zw^{-1}$ have a common zero at $(1, -1)$.

However, if instead we chose $p = t = 1$ $q = 2, r = 0$ we now have that $a_{q,\,t}(z, w) \equiv 1$ and hence condition (ii) is satisfied as well. In terms of $\Phi(a)$ and $\Phi(b)$ we have

$$\Phi \frac{b(z, w)}{a(z, w)} = \Phi \frac{(zw^{-2})^2 w^2 + (zw^{-2})w + w}{1 + (zw^{-2})w + w} = \frac{\alpha^2 \beta^2 + \alpha\beta + \beta}{1 + \alpha\beta + \beta}.$$

Now it is easily checked that $\alpha^2\beta^2 + \alpha\beta + \beta$ and $1 + \alpha\beta + \beta$ only have a common zero at $(\alpha_0, \beta_0) = (-1/2, -2)$.

We conclude that although $[b(z, w)]/[a(z, w)]$ is not stabilizable in $R_{1, 0, 1, 1}$ it is stabilizable in $R_{1, 0, 2, 1}$.

In order to draw conclusions for the general case we first prove the following lemma.

(3.137). LEMMA: *Let*

$$p, q, r, t \epsilon \mathbb{Z}_+$$

be such that

$$pt - qr = 1.$$

Let

$$(x_1, y_1), \ldots (x_n, y_n)$$

be a finite set of points in

$$Q_1 \setminus \{(0, 0), (1, 1)\} \ (Q_1 \text{ is the first quadrant of } \mathbb{R}^2) \text{ such that}$$
for all $i = 1, \ldots n$,

$$x_i^t \le y_i^q \text{ and } y_i^p \le x_i^r.$$

Then

$$\exists p', q', r', t' \ \epsilon \mathbb{Z}_+ \quad \text{with} \quad p't' - q'r' = 1$$

and

$$H_{p, r} \cap H_{q, t} \subset H_{p', r'} \cap H_{q' t'}$$

such that for each i, $i = 1, \ldots n$ *either*

$$x_i^{t'} > y_i^{q'} \quad \text{or} \quad y_i^{p'} > x_i^{r'}$$

Proof. We first note that

$$x_i^t = y_i^q \qquad y_i^p = x_i^r$$

implies either $(x_i, y_i) = (0, 0)$ or $(1, 1)$.

But we have excluded $(0, 0)$ and $(1, 1)$ from our values for (x_i, y_i). Therefore we can split the (x_i, y_i)'s into two sets:

$$Z_1 = \{(x_i, y_i): \qquad x_i^t < y_i^q, y_i^p \le x_i^r\}$$

and

$$Z_2 = \{(x_i, y_i): \quad x_i^t = y_i^q, \; y_i^p < x_i^r\}.$$

Let

$$I_j = \{i \in \{1, \ldots n\}: \quad (x_i, y_i) \in Z_j\} \quad j = 1, 2.$$

(Then

$$\{1, \ldots, n\} = I_1 U I_2$$

and $I_1 \cap I_2$ is empty).

We can assume that q and r are not zero since for example if $q = 0$, then $p = t = 1$ and we define a new (larger) causality cone $H_{p_1, r_1} \cap H_{q_1, t_1}$ by

$$p_1 = p \qquad t_1 = t + r$$

$$q_1 = p \qquad r_1 = r.$$

Since p, t can never be zero we therefore assume that p, t, q, r are all greater than zero.

For $i \in I_1$, $x_i^t < y_i^q$ and therefore

$$\left(\frac{y_i^q}{x_i^t}\right)^k \to \infty \quad \text{as} \quad k \to \infty.$$

(Note that x_i cannot be zero since this would imply $y_i = 0$ contradicting $x_i^t < y_i^q$).

It follows that for each $i \in I_1$ there exists $k_i \in \mathbb{Z}_+$ such that

$$y_i^{p + k_i q} = y_i^p \, y_i^{k_i q} > x_i^r \, x_i^{k_i t} = x_i^{r + k_i t}.$$

Now let

$$k = \max_{i \in I_1} k_i$$

$$p' = p + kq$$

$$r' = r + kt$$

then

$$p't - qr' = pt - qr = 1$$

and

for $i \epsilon I_1$

$$y_i^{p'} > x_i'. \tag{3.138}$$

Certainly $H_{p,\,r} \cap H_{q,\,t} \subset H_{p',\,r'} \cap H_{q,\,t}$ for if x, $y \epsilon R^2$ is such that

$$px + ry \geq 0, \qquad qx + ty \geq 0$$

then

$$p'x + r'y = px + ry + k(qx + ty) \geq 0.$$

Now consider $(x_i,\ y_i) \epsilon Z_2$.
Since

$$x_i^t = y_i^q$$

we have

$$y_i^{p'} = y_i^p\, y_i^{qk} = y_i^p\, x_i^{tk} < x_i'\, x_i^{tk} = x_i'$$

i.e. for $(x_i,\ y_i) \epsilon Z_2$ we have

$$x_i^t = y_i^q \qquad y_i^{p'} < x_i'.$$

Therefore for $i \epsilon I_2$

$$x_i^{t+r'} = x_i^t\, x_i' > x_i^t\, y_i^{p'} = y_i^q\, y_i^{p'} = y_i^{q+p'}.$$

So if we let

$$t' = t + r', \qquad q' = q + p'$$

then

$$p't' - q'r' = p't - qr' = 1$$

and

$$x_i' > y_i^{q'} \qquad i \epsilon I_2. \tag{3.139}$$

Also by an argument similar to that given before

$$H_{p,\,r} \cap H_{q,\,t} \subset H_{p',\,r'} \cap H_{q,\,t} \subset H_{p',\,r'} \cap H_{q',\,t'}$$

and so by (3.138) and (3.139), the proof of the lemma is complete.
We can now state and prove the main theorem of this chapter.

(3.140). THEOREM. *Let*

$$\frac{b(z, w)}{a(z, w)}$$

be an irreducible representation of $p(z, w) \epsilon \bar{R}_{p, r, q, t}$. *Then there exists a causality cone* $C_c^* = H_{p^*, r^*} \cap H_{q^*, t^*}$ *such that*

$$C_c \triangleq H_{p, r} \cap H_{q, t} \subset C_c^*$$

and such that $p(z, w)$ *is stabilizable by means of a compensator*

$$c(z, w) \epsilon R_{p^*, r^*, q^*, t^*}$$

if and only if

$$b(z, w) \quad and \quad a(z, w)$$

have no common zeroes on T^2.

Proof. If $b(z, w)$, $a(z, w)$ have a common zero $(z_0, w_0) \epsilon T^2$, then in whatever causality cone $C_c = H_{p, r} \cap H_{q, t}$ the supports of $a(z, w)$ and $b(z, w)$ are considered to be, $\Phi(b) (\alpha, \beta)$ and $\Phi(a) (\alpha, \beta)$ have a common zero at $(\alpha_0, \beta_0) \epsilon T^2$
where

$$\alpha_0 = \frac{z_0^{t^*}}{w_0^{q^*}}. \qquad \beta_0 = \frac{w_0^{p^*}}{z_0^{r^*}}.$$

Conversely, suppose $a(z, w)$ and $b(z, w)$ have no common zero on T^2.

Since $[b(z, w)]/[a(z, w)]$ is irreducible $b(z, w)$ and $a(z, w)$ have only finitely many zeroes in common. (If a, b involve powers of z^{-1} and w^{-1} these zeroes are found by multiplying $b(z, w)$ and $a(z, w)$ to make them polynomials and then finding those common zeroes (z_0, w_0) of these polynomials for which $z_0 \neq 0$ and $w_0 \neq 0$. If a, b involve powers of z^{-1} but not w^{-1} just multiplying each by an appropriate power of z and find zeroes (z_0, w_0) for which $z_0 \neq 0$; similarly if a, b involve powers of w^{-1} but not z^{-1}.)

Let the common zeroes of $a(z, w)$ and $b(z, w)$ be

$$(z_1, w_1), (z_2, w_2), \ldots, (z_m, w_m)$$

and let the labelling be such that

$$(z_i, w_i) \epsilon V(p, r, q, t) \qquad i = 1, \ldots n$$
$$(z_i, w_i) \notin V(p, r, q, t) \qquad i = n + 1, \ldots m.$$

Note that $(0, 0)$ cannot be a common zero of $a(z, w)$, $b(z, w)$. Indeed, the only time this could happen is when $a(z, w)$ and $b(z, w)$ are polynomials but we have assumed $a(z, w)$ has non-zero constant term.

Now by Lemma (3.137), there exists a causality cone $C'_c = H_{p', r} \cap H_{q', t'}$ such that

$$H_{p, r} \cap H_{q, t} \subset C'_c$$

and such that

$$(z_i, w_i) \notin V(p', r', q', t') \qquad \text{for } i = 1, \ldots n. \qquad (3.141)$$

Now it is easy to show that

$$V_{p', r', q', t'} \subset V_{p, r, q, t}$$

so from (3.141) we have that

$$(z_i, w_i) \notin V(p', r', q', t') \qquad \text{for } i = 1, \ldots m.$$

In other words, $b(z, w)$ and $a(z, w)$ satisfy condition (i) of (3.135) (with p', r', q', t' in place of p, r, q, t).

To satisfy (3.135) (ii) and (iii) let

$$q^* = q' + p', \qquad t^* = t' + r'$$
$$p^* = p' + q^*, \qquad r^* = r' + t^*.$$

Then

$$p^*t^* - q^*r^* = 1$$
$$H_{p, r} \cap H_{q, t} \subset H_{p', r'} \cap H_{q', t'} \subset H_{p^*, r^*} \cap H_{q^*, t^*}$$

and

$$a_{q^*, r^*}(z, w) = a_{p^*, r^*}(z, w) = \text{const.} \neq 0.$$

So (i), (ii), and (iii) are satisfied (with p^*, r^*, q^*, t^* replacing p, r, q, t) and therefore $[b(z, w)]/[a(z, w)]$ is stabilizable in R_{p^*, r^*, q^*, t^*}.

(3.142). NOTE. The construction in the proof of Theorem (3.140) may not in practice be the best way of finding p^*, r^*, q^*, t^*.

(3.143). EXAMPLE.

$$\frac{b(z, w)}{a(z, w)} = \frac{z + w}{1 - z + w}.$$

The only common zero of a and b is at $(\frac{1}{2}, -\frac{1}{2})$.
Take

$$p^* = t^* = 1, \qquad q^* = 0, \qquad r^* = 2.$$

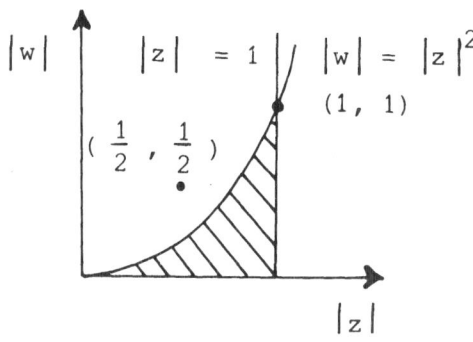

Fig. 3.3   The shaded region shows the region $\{(|z|, |w|): (z, w)\epsilon V(1, 2, 0, 1)\}$.

Then (see Figure 3.3), $a(z, w)$ and $b(z, w)$ have no common zeroes in $V(1, 2, 0, 1)$ and so (3.135) is satisfied (conditions (ii) and (iii) are easily checked).
Hence

$$\frac{z + w}{1 - z + w} \qquad \text{is stabilizable in } R_{1, 2, 0, 1}.$$

If we wish to stabilize it we first map $R_{1, 2, 0, 1}$ to $R_{1, 0, 0, 1}$ under Φ. So

$$\Phi\left(\frac{z + w}{1 - z + w}\right) = \Phi\left(\frac{z + (wz^{-2})z^2}{1 - z + (wz^{-2})z^2}\right) = \frac{\alpha + \beta\alpha^2}{1 - \alpha + \beta\alpha^2} =$$

$$\frac{b_0(\alpha, \beta)}{a_0(\alpha, \beta)}.$$

We then stabilize $[b_0(\alpha, \beta)]/[a_0(\alpha, \beta)]$ using the methods of Section 3.5 to obtain a causal compensator $c_1(\alpha, \beta)$. Then $\Phi^{-1}: R_{1, 0, 0, 1} \rightarrow R_{1, 2, 0, 1}$ will map $c_1(\alpha, \beta)$ to a $c(z, w)\epsilon R_{1, 2, 0, 1}$ which will stabilize $[b(z, w)]/[a(z, w)]$.

(3.144). EXAMPLE.

$$\frac{b(z, w)}{a(z, w)} = \frac{z^2 w^{-2} + zw^{-1} + w + 4z}{1 + zw^{-1} + 2w} \; \epsilon \bar{R}_{1, 0, 1, 1}$$

$a(z, w)$ and $b(z, w)$ have just one common zero, occurring at $(\frac{1}{8}, -\frac{1}{4})$.

$a_{q, t}(z, w)$ and $b_{q, t}(z, w)$ also have a common zero, occurring at $zw^{-1} = -1$.

Let

$$p_1 = p + q = 2, \qquad q_1 = q = 1$$
$$r_1 = t = 1, \qquad\qquad t_1 = t = 1.$$

Then

$$p_1 r_1 - q_1 t_1 = 1.$$

Now let

$$p' = p_1 = 2 \qquad q' = q_1 + 2p_1 = 5$$
$$r' = r_1 = 1 \qquad t' = t_1 + 2r_1 = 3.$$

Then

$$p't' - q'r' = 1, \qquad |\tfrac{1}{8}|^3 > |-\tfrac{1}{4}|^5$$

and

$$a_{q', t'}(z, w) = 1.$$

So

$$\frac{z^2 w^{-2} + zw^{-1} + w + 4z}{1 + zw^{-1} + 2w} \quad \text{is stabilizable in}$$

$$R_{2, 1, 5, 3}.$$

3.10. STABILIZATION OF MIMO WEAKLY CAUSAL SYSTEMS

$$\Phi : S_{p, r, q, t} \rightarrow S_{1, 0, 0, 1}$$

induces a map

$$\Phi : S_{p, r, q, t}^{m \times k} \rightarrow S_{1, 0, 0, 1}^{m \times k}$$

where for

$$H = [h_{ij}]_{m \times k} \epsilon S^{m \times k}_{p, r, q, t}$$

$$\Phi(H) = [\Phi(h_{ij})]_{m \times k} \epsilon S^{m \times k}_{1, 0, 0, 1}. \tag{3.145}$$

Φ also restricts obviously to mappings

$$\text{from} \quad R^{m \times k}_{p, r, q, t} \quad \text{to} \quad R^{m \times k}_{1, 0, 0, 1},$$

$$\text{from} \quad \bar{R}^{m \times k}_{p, r, q, t} \quad \text{to} \quad \bar{R}^{m \times k}_{1, 0, 0, 1},$$

and

$$\text{from} \quad P^{m \times k}_{p, r, q, t} \quad \text{to} \quad P^{m \times k}_{1, 0, 0, 1}.$$

Let

$$P(z, w) \epsilon R^{m \times k}_{p, r, q, t},$$

$$P = [p_{ij}]_{m \times k}$$

where

$$p_{ij}(z, w) = \frac{b_{ij}(z, w)}{a_{ij}(z, w)} \epsilon R_{p, r, q, t} \tag{3.146}$$

is irreducible $i = 1, \ldots m, \qquad j = 1, \ldots, k$

and

$$b_{ij}(z, w), a_{ij}(z, w) \epsilon P_{p, r, q, t} \tag{3.147}$$

with $a_{ij}(z, w)$ having non-zero constant term $i = 1, \ldots m,$
$j = 1, \ldots k.$ Let

$$\phi(a_{ij}(z, w)) = d_{ij}(\alpha, \beta) \tag{3.148}$$

$$\phi(b_{ij}(z, w)) = n_{ij}(\alpha, \beta) \qquad i = 1, \ldots m, j = 1, \ldots k.$$

Then

$$\Phi(P(z, w)) = H(\alpha, \beta) = \left[\frac{n_{ij}(\alpha, \beta)}{d_{ij}(\alpha, \beta)} \right] \tag{3.149}$$

where

$$\frac{n_{ij}(z, w)}{d_{ij}(z, w)} \quad \text{is irreducible} \qquad i = 1, \ldots m, j = 1, \ldots k.$$

$$\tag{3.150}$$

(3.151). A representation of P of the form $P = A^{-1}B$ where

$$A \epsilon P_{p,\,r,\,q,\,t}^{m \times m} \quad \text{and} \quad B \epsilon P_{p,\,r,\,q,\,t}^{m \times k}$$

is called a left matrix fraction description (LMFD) of P over $P_{p,\,r,\,q,\,t}$.
Similarly one can define an RMFD for P over $P_{p,\,r,\,q,\,t}$.

(3.152). An LMFD $P = A^{-1}B$ of P over $P_{p,\,r,\,q,\,t}$ is called irreducible if any left common factor of A and B over $P_{p,\,r,\,q,\,t}$ has as determinant a constant. (A left common factor of A and B over $P_{p,\,r,\,q,\,t}$ is a matrix $L \epsilon P_{p,\,r,\,q,\,t}^{m \times m}$ such that

$$A = LA_1, \quad B = LB_1 \quad \text{for some}$$

$$A_1 \epsilon P_{p,\,r,\,q,\,t}^{m \times m}, \quad B_1 \epsilon P_{p,\,r,\,q,\,t}^{m \times k}.$$

Because of the ring isomorphism properties of Φ: $R_{p,\,r,\,q,\,t} \rightarrow R_{1,\,0,\,0,\,1}$ the correspondence

$$A^{-1}B \leftrightarrow (\Phi(A))^{-1}\,\Phi(B) \tag{3.153}$$

defines a bijective correspondence between (irreducible) LMFD's of P over $P_{p,\,r,\,q,\,t}$ and (irreducible) LMFD's of $H = \Phi(P)$ over $P_{1,\,0,\,0,\,1}$.

Since from (3.147) we are assuming that the $a_{ij}(z, w)$'s have non-zero constant term, then any irreducible representation $A^{-1}B$ will have the property that

$$\det(A)(z, w) \tag{3.154}$$

has a non-zero constant term. It then follows (see proof of (3.43) and (3.46)) that if

$$A \epsilon R_{p,\,r,\,q,\,t}^{m \times m} \cap R_{p',\,r',\,q',\,t'}^{m \times m}$$

and

$$B \epsilon R_{p,\,r,\,q,\,t}^{k \times m} \cap R_{p',\,r',\,q',\,t'}^{k \times m} \quad \text{then}$$

(3.155). $A^{-1}B$ is irreducible in $R_{p,\,r,\,q,\,t}$ if and only if it is irreducible in $R_{p',\,r',\,q',\,t'}$.

We have, as in the causal case, that if $A^{-1}B$ is an irreducible LMFD of P over $P_{p,\,r,\,q,\,t}$ then

(3.156). $P \epsilon \bar{R}_{p,\,r,\,q,\,t}^{m \times k}$ if and only if all entries of B have zero constant term.
Due to the isomorphism Φ, various other properties of LMFD's over

$P_{1, 0, 0, 1}$ go through for LMFD's over $P_{p, r, q, t}$ but these will not be of interest to us here.

Matrix fraction descriptions of elements of $R^{m \times k}_{p, r, q, t}$ and $\bar{R}^{m \times k}_{p, r, q, t}$ allow us very easily to derive the matrix counterparts of the scalar results of Section 3.9.

Let

$$P \epsilon \bar{R}^{m \times k}_{p, r, q, t}$$

and let

$$P = A_L^{-1} B_L$$

be an irreducible LMFD of P over $P_{p, r, q, t}$. Then from (3.154) and (3.155) it is clear that,

$$\det(A_L) \text{ has a non-zero constant term and}$$

$$\text{the entries of } B_L \text{ have zero constant term.}$$

If $d(z, w)$ is an mth order minor of $[A_L B_L]$ we denote by $d_{p, r}(z, w)$ that portion of $d(z, w)$ involving only powers of $w^p z^{-r}$ (including the constant term) and by $d_{q, t}(z, w)$ that portion of $d(z, w)$ involving only powers of $z^t w^{-q}$ (including the constant term).

Then the matrix counterpart of (3.133) is

(3.157). PROPOSITION. *$P = A_L^{-1} B_L$ satisfying (3.155) and (3.156) is stabilizable by a compensator in $R^{k \times m}_{p, r, q, t}$, if and only if*

(i) *the mth order minors of $[A_L B_L]$ have no common zeroes in $V(p, r, q, t)$*

(ii) *the polynomials in $z^t w^{-q}$, $d^{(i)}_{q, t}(z, w)$, have no common zero in $|z^t w^{-q}| \leq 1$. (where $d^{(i)}(z, w)$ $i = 1, \ldots \binom{m + k}{m}$ are the mth order minors of $[A_L B_L]$)*

(iii) *the polynomials in $w^p z^{-r}$, $d^{(i)}_{p, r}(z, w)$, have no common zero in $|w^p z^{-r}| \leq 1$.*

Proof. Follows from the ring isomorphism $\Phi: R_{p, r, q, t} \rightarrow R_{1, 0, 0, 1}$ and the results of Section 3.5.

The matrix version of Theorem (3.138) is

(3.158). THEOREM. *Let*

$$P = A_L^{-1} B_L \epsilon \bar{R}_{p, r, q, t}$$

satisfy (3.155) *and* (3.156). *Then there exists a causality cone*

$$C_c^* = H_{p^*, r^*} \cap H_{q^*, t^*}$$

such that

$$H_{p, r} \cap H_{q, t} \subset C_c^*$$

and such that P is stabilizable by means of a compensator

$$C(z, w) \epsilon R_{p^*, r^*, q^*, t^*}^{k \times m}$$

if and only if the mth order minors $[A_L B_L]$ have no common zeroes on T^2.

Proof. A common zero $(z_0, w_0) \epsilon T^2$ of the mth order minors would imply that whatever causality cone we choose, $(\Phi(A))^{-1}\Phi(B)$ will be an irreducible representation of $\Phi(P)$ such that m^{th} order minors of

$$[\Phi(A)(\alpha, \beta) \ \Phi(B)(\alpha, \beta)]$$

have a common zero on T^2 and $\Phi(P)$ is therefore not stabilizable by a causal system (3.112).

Conversely, because $A_L^{-1}B_L$ is irreducible, the mth order minors of $[A_L B_L]$ have only a finite number of common zeroes which, by hypothesis, do not lie on T^2. Also as in the scalar case, $(0, 0)$ cannot be one of these common zeroes. We can then (following the same proof as in the scalar case) find p^*, r^*, q^*, t^* such that $H_{p, r} \cap H_{q, t} \subset H_{p^*, r^*} \cap H_{q^*, t^*}$ and such that (i), (ii), and (iii) of (3.159) are satisfied.

As in the scalar case once we have p^*, r^*, q^*, t^* as in the above theorem, the compensators can be obtained by mapping R_{p^*, r^*, q^*, t^*} to $R_{1, 0, 0, 1}$, using the causal techniques of Section 3.5, and then mapping back again.

3.11. CONCLUSIONS

The general problem investigated in this chapter is the structural stability of 2–D feedback systems where the plant and compensator each correspond to discrete 2–D causal or weakly causal multi-input/multi-output systems. In particular, necessary and sufficient conditions are obtained for an unstable plant to be stabilizable and a classification of the stabilizing compensators is given. The recent factorization results for bivariate polynomial matrices with coefficients over an arbitrary but fixed field [3.8], which enable us to obtain irreducible matrix fraction descriptions of bivariate rational transfer matrices, proved to be very useful in this study.

A compensator required to stabilize an unstable plant (provided of course this plant is stabilizable), may be constructed using the techniques described in Chapter 6. Also, the contents of Chapter 7 dealing with a derivation of sufficient conditions for the solution of a system of linear equations over a bivariate polynomial ring are useful in the design of compensators to provide 2–D feedback systems with certain desirable properties. This chapter should motivate further research, especially because multidimensional feedback systems have been proposed for various purposes like iterative image processing and image restoration [3.11], [3.12]. Such image processing systems that contain feedback loops are sometimes known to oscillate in space and time [3.12] and these undesirable oscillations can only be avoided if proper stability conditions are imposed on the feedback systems. In the design of 2–D feedback systems, the possibility of incorporating the weakly causal property in the design of recursive compensators broadens considerably the class of plants that can be stabilized (see examples (3.176) and (3.143) in this chapter).

REFERENCES

[3.1]   R. Eising, 'State Space Realization and Inversion of 2–D Systems', *IEEE Trans. Circuits and Systems*, Vol. CAS–27, No. 7, pp. 612–619, July 1980.

[3.2]   B. O'Connor and T. Huang, 'Stability of General Two Dimensional Recursive Digital Filters', *IEEE Trans. Acoust., Speech, Signal Processing*, Vol. ASSP–26, pp. 550–560, December 1978.

[3.3]   R. Eising, 'Realization and Stabilization of 2–D Systems', *IEEE Trans. Automatic Control*, Vol. AC–23, No. 5, pp. 793–799, October 1978.

[3.4]   J. Cadzow, '*Discrete-Time Systems: An Introduction with Interdisciplinary Applications*', Prentice Hall, NJ, 1973.

[3.5]   C. Desoer, R-W. Liu, J. Murray, and R. Saeks, 'Feedback System Design: The Fractional Representation Approach to Analysis and Synthesis', *IEEE Trans. Automatic Control*, Vol. AC–25, No. 3, pp. 399–412, June 1980.

[3.6]   C. Desoer and W. Chan, 'The Feedback Interconnection of Lumped Linear Time-invariant Systems', *Journal of the Franklin Institute*, Vol. 300, No. 5/6, pp. 335–351, Nov./Dec. 1975.

[3.7]   M. Vidyasagar, 'On the Use of Right-Coprime Factorizations in Distributed Feedback Systems Containing Unstable Subsystems', *IEEE Trans. Circuits and Systems*, Vol. CAS–25, No. 11, pp. 916–921, November 1978.

[3.8]   J. P. Guiver and N. K. Bose, 'Polynomial Matrix Primitive Factorization Over Arbitrary Coefficient Field and Related Results', *IEEE Trans. Circuits and Systems*, Vol. CAS–30, No. 10, October 1982.

[3.9]   D. Youla and G. Gnavi, 'Notes on *n*–Dimensional System Theory', *IEEE Trans. Circuits and Systems*, Vol. CAS–26, No. 2, pp. 105–111, February 1979.

[3.10] M. Morf, B. Lévy, and S-Y. Kung, 'New Results on 2–D Systems Theory, Part I: 2–D Polynomial Matrices, Factorization and Coprimeness', *Proc. IEEE*, Vol. 65, pp. 861–872, June 1977.

[3.11] G. Ferrani and G. Häusler, 'TV Optical Feedback Systems', *Optical Engineering*, 19, July/August 1980, pp. 442–451.

[3.12] G. Häusler and N. Streibl, 'Stability of Spatio-temporal Feedback Systems', *Optica Acta*, 1930, 1983, pp. 171–187.

Chapter 4

E. W. Kamen

Stabilization of Linear Spatially-Distributed Continuous-Time and Discrete-Time Systems

4.1. INTRODUCTION

In this chapter we study a class of multidimensional systems whose inputs and outputs are functions of a time variable and a discrete spatial variable. In particular, we shall concentrate on linear spatially-distributed time-invariant systems. In the single-input single-output continuous-time case, such a system can be represented by the input/output relationship

$$y(t, r) = \int_{-\infty}^{t} \left[\sum_{j=-\infty}^{\infty} w(t - \tau, r - j)u(\tau, j) \right] d\tau. \qquad (4.1)$$

In (4.1), $u(t, r) \epsilon \mathbb{R}$ (reals) is the input or control at time $t \epsilon \mathbb{R}$ and spatial point $r \epsilon \mathbb{Z}$ (the integers), $y(t, r) \epsilon \mathbb{R}$ is the output at time t and point r, and $w(t, r) \epsilon \mathbb{R}$ is the value of the impulse response at time t and point r. We assume that $u(t, r)$ and $w(t, r)$ are constrained so that the bi-infinite sum and integral in (4.1) are well defined. We shall make this precise shortly. Note that the upper limit in the integral in (4.1) is t since we are assuming causality in time; that is, $w(t, r) = 0$ for all $t < 0$. We are *not* assuming a one-sided causality in space; that is, $w(t, r) \neq 0$ in general for all $r \epsilon \mathbb{Z}$.

Systems that can be represented by an input/output relationship of the form (4.1) arise in the study of long strings of coupled continuous-time systems, such as strings of vehicles (see Melzer and Kuo [4.1] and Chu [4.2]). They also result from the spatial discretization of partial differential equations which are functions of time and one spatial coordinate. An example of such a discretization for a long seismic cable will be given later.

In the application to long strings of coupled systems or discretized partial differential equations, there is usually some damping along the distributed structure, with the result that we can make the following summability assumption:

N. K. Bose (ed.), Multidimensional Systems Theory, 101–146
© 1985 by D. Reidel Publishing Company.

$$\sum_{j=-\infty}^{\infty} |w(t, j)| < \infty \quad \text{for all} \quad t > 0. \tag{4.2}$$

If (4.2) is satisfied and if the input $u(t, r)$ is a bounded function of the spatial variable r for every $t \in \mathbb{R}$, from (4.1) we have that the output response $y(t, r)$ resulting from $u(t, r)$ is also a bounded function of r for every $t \in \mathbb{R}$. Furthermore, we have that

$$|y(t, r)| \leq \int_{-\infty}^{t} \left[\sum_{j=-\infty}^{\infty} |w(t - \tau, r - j)| \right] \times \left[\sup_{j \in \mathbb{Z}} |u(\tau, j)| \right] d\tau. \tag{4.3}$$

From (4.3) we see that if $\Sigma_j |w(t, j)|$ and $\sup_r |u(t, r)|$ are locally integrable functions of t (with the support of $\sup_r |u(t, r)|$ bounded on the left), then the expression (4.1) for $y(t, r)$ is well defined. In addition, under the conditions given above, we can interchange the summation and integral in (4.1) so that we have

$$y(t, r) = \sum_{j=-\infty}^{\infty} \left[\int_{-\infty}^{t} w(t - \tau, r - j) u(\tau, j) \, d\tau \right]. \tag{4.4}$$

The proof that (4.1) and (4.4) are equivalent follows from the property that the sequence of partial sums for $\Sigma_j w(t, j)$ converges uniformly in t for t belonging to any compact subset of $\mathbb{R}$.

The system with input/output relationship (4.1) or (4.4) is both time-invariant and spatially invariant. In other words, if $y(t, r)$ is the output response resulting from input $u(t, r)$, then $y(t - \tau, r - j)$ is the output response resulting from input $u(t - \tau, r - j)$ for any $\tau \in \mathbb{R}$ and any $j \in \mathbb{Z}$. A spatially-varying (but time-invariant) system can be represented by the input/output relationship

$$y(t, r) = \int_{-\infty}^{t} \left[\sum_{j=-\infty}^{\infty} w(t - \tau, r, j) u(\tau, j) \right] d\tau,$$

where now the impulse response function $w(t, r, j)$ is a real-valued function defined on the Cartesian product $\mathbb{R} \times \mathbb{Z} \times \mathbb{Z}$. (Here we are assuming that the system is strictly causal; i.e., $w(t, r, j)$ does not contain an impulse at $t = 0$.) In this chapter we shall restrict our attention to the spatially-invariant case.

The transfer-function representation for a system given by (4.1) is constructed in terms of the joint Laplace/z transform which is defined as follows. Let f be a real-valued function defined on the Cartesian product

$\mathbb{R} \times \mathbb{Z}$ with $f(t, r) = 0$ for all $t < t_f$ (i.e., f is one-sided in time). The joint Laplace/z transform $F(s, z)$ of f is defined by

$$F(s, z) = \int_{-\infty}^{\infty} \left[\sum_{j=-\infty}^{\infty} f(t, j)z^{-j} \right] e^{-st} \, dt. \tag{4.5}$$

We assume that the function f is constrained so that $F(s, z)$ is well defined for (s, z) belonging to some subset of $\mathbb{C} \times \mathbb{C}$ where $\mathbb{C}$ = field of complex numbers. For example, suppose that there exist real numbers σ and a with $a > 0$, such that

$$\sum_{j=-\infty}^{\infty} |f(t, j)||a|^{-j} = g(t) < \infty, \quad \text{all} \quad t \in \mathbb{R}, \tag{4.6a}$$

and

$$\int_{-\infty}^{\infty} g(t)e^{-\sigma t} \, dt < \infty. \tag{4.6b}$$

Then the joint Laplace/z transform $F(s, z)$ is well defined for all (s, z) such that $\text{Re } s \geq \sigma$ and $|z| = a$. We should note that the conditions (4.6a, b) are not necessary for the existence of the transform.

The joint Laplace/z transform has all the expected properties. For instance, suppose that we have two functions $f(t, r)$ and $\gamma(t, r)$, both of which satisfy the properties (4.6a, b) for some σ, a. Then defining the joint convolution

$$(f *_t *_r \gamma)(t, r) = \int_{-\infty}^{\infty} \left[\sum_{j=-\infty}^{\infty} f(t - \tau, r - j)\gamma(\tau, j) \right] d\tau,$$

the transform of $f *_t *_r \gamma$ is equal to $F(s, z)\Gamma(s, z)$, where $F(s, z)$ ($\Gamma(s, z)$) is the transform of $f(t, r)$ ($\gamma(t, r)$).

Now consider the system with the input/output relationship (4.1). With the assumption (4.2), we see that the impulse response function $w(t, r)$ has a joint Laplace/z transform $W(s, z)$ if

$$\int_{0}^{\infty} \left[\sum_{j=-\infty}^{\infty} |w(t, j)| \right] e^{-\sigma t} \, dt < \infty \quad \text{for some} \quad \sigma \in \mathbb{R},$$

in which case, $W(s, z)$ is well defined for all (s, z) such that $\text{Re } s \geq \sigma$, $|z| = 1$. Thus if the input $u(t, r)$ has the properties (4.6a, b) for some σ_u, and $a = 1$, we can take the joint Laplace/z transform of both sides of (4.1) which gives

$$Y(s, z) = W(s, z)U(s, z), \tag{4.7}$$

where $Y(s, z)$ is the transform of the output response $y(t, r)$. The relationship (4.7) is the *transfer-function representation* of the system, and $W(s, z)$ is the *transfer function* of the system. Since $W(s, z)$ is well defined when $|z| = 1$ (i.e., $z = e^{i\omega}$ where $i = \sqrt{-1}$ and $\omega \in [0, 2\pi]$), it turns out that the system is completely characterized by

$$W(s, \omega) = W(s, z)|_{z = \exp(i\omega)}.$$

We shall consider this later.

EXAMPLE 4.1   From the work of El-Sayed and Krishnaprasad [4.3], a long seismic cable used in offshore oil exploration can be modeled by the input/output differential equation

$$\frac{d^2y(t, r)}{dt^2} + (c/M)\frac{dy(t, r)}{dt} + (k/M)y(t, r) - \tag{4.8}$$
$$(k/2M)[y(t, r - 1) + y(t, r + 1)] =$$
$$(1/M)u(t, r).$$

In (4.8), $y(t, r)$ is the position at time t of the rth cable segment relative to some reference, $u(t, r)$ is the input or control at time t applied to the rth cable segment, M is the mass of a cable segment, c is a damping coefficient, and k is a fixed positive constant. The differential equation (4.8) results from the spatial discretization of a partial differential equation representation of the cable (see [4.3] for details on the discretization technique). Assuming the initial conditions

$$y(0, r) = 0 \quad \text{for all} \quad r \in \mathbb{Z},$$

and

$$\left.\frac{dy(t, r)}{dt}\right|_{t = 0} = 0 \quad \text{for all} \quad r \in \mathbb{Z},$$

if we take the joint Laplace/z transform of both sides of (4.8), we get

$$s^2Y(s, z) + (c/M)sY(s, z) + (k/M)Y(s, z) -$$
$$(k/2M)[z^{-1}Y(s, z) + zY(s, z)] = (1/M)U(s, z).$$

Solving for $Y(s, z)$, we have that

$$Y(s, z) = W(s, z)U(s, z),$$

where $W(s, z)$ is the transfer function of the cable given by

$$W(s, z) = \frac{1/M}{s^2 + (c/M)s + (k/M)[1 - 0.5z^{-1} - 0.5z]}. \quad (4.9)$$

Note that in this example the transfer function $W(s, z)$ is a rational function of both s and z. We shall conclude the example by computing the impulse response function $w(t, r)$ for the cable. We can compute $w(t, r)$ by taking the inverse Laplace/Fourier transform of $W(s, \omega) = W(s, z)|_{z = \exp(i\omega)}$. Setting $z = \exp(i\omega)$ in (4.9), we have

$$W(s, \omega) = \frac{1/M}{s^2 + (c/M)s + (k/M)(1 - \cos \omega)}.$$

Define $\beta(\omega) = 0.5[c^2/M^2 - (4k/M)(1 - \cos \omega)]^{1/2}$, so that

$$s^2 + (c/M)s + (k/M)(1 - \cos \omega) =$$
$$(s + c/2M + \beta(\omega))(s + c/2M - \beta(\omega)).$$

Then if $\beta(\omega) \neq 0$ for all $\omega \in [0, 2\pi]$, the poles of $W(s, \omega)$ are distinct for all $\omega \in [0, 2\pi]$, and $W(s, \omega)$ has the partial-fraction expansion

$$W(s, \omega) = \frac{1}{M\beta(\omega)}\left[\frac{1}{s + c/2M - \beta(\omega)} - \frac{1}{s + c/2M + \beta(\omega)}\right].$$

Taking the inverse Laplace/Fourier transform of $W(s, \omega)$, we get that

$$w(t, r) = (1/2\pi M)e^{-(c/2M)t} \int_0^{2\pi} \frac{1}{\beta(\omega)}\left[e^{\beta(\omega)t} - e^{-\beta(\omega)t}\right]e^{i\omega r} \, d\omega.$$

It does not appear to be possible to write $w(t, r)$ in closed form. However, since $1/\beta(\omega)$ and $\exp\beta(\omega)t$ are continuously differentiable functions of ω, it is possible to conclude that $w(t, r)$ is an absolutely summable function of r for all t. Thus, the impulse response function for the cable does satisfy the summability condition (4.2). Finally, we note that the above expression for $w(t, r)$ could be computed using the fast Fourier transform (see Section 4.3).

In this chapter we shall also consider linear spatially-distributed time-invariant discrete-time systems. In the single-input single-output spatially-invariant case, such a system can be represented by the input/output convolution sum

$$y(k, r) = \sum_{j_1 = -\infty}^{k} \sum_{j_2 = -\infty}^{\infty} w(k - j_1, r - j_2)u(j_1, j_2). \quad (4.10)$$

In (4.10), the input $u(k, r)$, output $y(k, r)$, and unit-pulse response

$w(k, r)$ are all real-valued functions defined on the Cartesian product $\mathbb{Z} \times \mathbb{Z}$. Here k is the discrete time variable and r is the discrete spatial variable. We are again assuming causality in time so that the sum on j_1 in (4.10) terminates at $j_1 = k$.

Systems given by (4.10) arise in the study of long strings of coupled discrete-time systems. They may also result from the discretization in time of a spatially-distributed continuous-time system. In Section 4.3 we study the problem of time discretization using a state model for the given system.

We should also point out that the relationship (4.10) could be the input/output representation of a half-plane causal shift-invariant two-dimensional digital filter. In this interpretation, both k and r are viewed as spatial variables.

The primary objective of this chapter is to study the existence and construction of stabilizing feedback compensators for spatially-distributed systems. More precisely, given the system defined by (4.1), we want to know if there is a feedback system defined by

$$\gamma(t, r) = \int_{-\infty}^{t} \left[\sum_{j=-\infty}^{\infty} w_{fb}(t - \tau, r - j)y(\tau, j) \right] d\tau,$$

such that taking the input $u(t, r)$ of the given system to be the output $\gamma(t, r)$ of the feedback system plus some possible external input $v(t, r)$, the resulting closed-loop system is *internally stable*. Here the notion of internal stability is defined in terms of state-space representations for the given system and for the feedback system. The state representation for multi-input multi-output spatially-distributed systems is formulated in the next section. The notion of stability is defined in Section 4.5, and in Sections 4.6–4.8, we give results on the existence and construction of stabilizing compensators. Criteria for stabilizability are specified in terms of both a state space representation and a transfer function matrix representation (with $z = e^{i\omega}$). Our approach is based on techniques from the theory of linear systems with coefficients in a commutative ring or algebra (e.g., see [4.4], [4.5], [4.6]).

The property of stabilizability by feedback is of fundamental importance in a wide range of control problems. For instance, various types of regulation problems such as tracking and disturbance rejection can be reduced to the problem of stabilizing an augmented system, a technique that is often employed in the control of finite-dimensional systems (i.e., systems with finite dimensional state space). In Section 4.9 we apply our stabilization results to the problem of tracking.

Finally, we wish to note that the techniques which are developed in this chapter can be extended to linear spatially-distributed systems with a two-dimensional discrete-spatial dependence. Such systems are actually three dimensional since there is one time variable and two spatial variables. Continuous-time systems with a two-dimensional discrete-spatial dependence may result from the spatial discretization of partial-differential equation models for systems that are distributed in a plane, such as large platforms in free space.

4.2. THE STATE REPRESENTATION AND INPUT/OUTPUT DESCRIPTION

In this section we consider both state representations and input/output descriptions for multi-input multi-output spatially-distributed systems.

4.2.1. *The State Representation*

Given a fixed positive integer n, let $\mathbb{R}^n$ denote the space of n-element column vectors with entries in the reals $\mathbb{R}$. Given a vector $x \in \mathbb{R}^n$, we denote the norm of x by $\|x\|$, where $\|x\|$ is any one of the usual norms on $\mathbb{R}^n$. The particular choice for $\|x\|$ is assumed to be fixed throughout the chapter.

Let $\ell^\infty(\mathbb{Z}, \mathbb{R}^n)$ denote the space of all functions v from $\mathbb{Z}$ into $\mathbb{R}^n$ which are bounded; that is,

$$\sup_{r \in \mathbb{Z}} \|v(r)\| < \infty.$$

It is well known that $\ell^\infty(\mathbb{Z}, \mathbb{R}^n)$ is a Banach space with the sup norm

$$\|v\|_\infty = \sup_{r \in \mathbb{Z}} \|v(r)\|.$$

Let $\ell^1(\mathbb{Z}, \mathbb{R})$ denote the space of all functions v from $\mathbb{Z}$ into $\mathbb{R}$ which are absolutely summable; i.e.,

$$\sum_{j=-\infty}^{\infty} |v(j)| < \infty.$$

The space $\ell^1(\mathbb{Z}, \mathbb{R})$ is also a Banach space with the ℓ^1 norm

$$\|v\|_1 = \sum_{j=-\infty}^{\infty} |v(j)|.$$

In addition, with the convolution operation

$$(v_1 * v_2)(r) = \sum_{j=-\infty}^{\infty} v_1(r - j)v_2(j),$$

$\ell^1(\mathbb{Z}, \mathbb{R})$ is a *commutative Banach algebra* with multiplicative identity equal to the unit pulse concentrated at the origin.

Given positive integers m and n, let P be an $n \times m$ matrix with entries in $\ell^1(\mathbb{Z}, \mathbb{R})$. Then P defines a bounded linear transformation from $\ell^\infty(\mathbb{Z}, \mathbb{R}^m)$ into $\ell^\infty(\mathbb{Z}, \mathbb{R}^n)$ given by

$$v \to P * v,$$

where

$$(P * v)(r) = \sum_{j=-\infty}^{\infty} P(r - j)v(j).$$

The proof that P is bounded follows by a standard argument which is omitted. Finally, we come to one of our central definitions.

DEFINITION 4.1. Let m, n, p be fixed positive integers. A state representation Σ over $\ell^1(\mathbb{Z}, \mathbb{R})$ is a quadruple $\Sigma = (F, G, H, J)$ of $n \times n$, $n \times m$, $p \times n$, $p \times m$ matrices with entries in $\ell^1(\mathbb{Z}, \mathbb{R})$.

A state representation $\Sigma = (F, G, H, J)$ defines a m-input p-output linear spatially-distributed continuous-time system given by the state equations

$$\frac{dx(t, r)}{dt} = \sum_{j=-\infty}^{\infty} F(r - j)x(t, j) + \sum_{j=-\infty}^{\infty} G(r - j)u(t, j) \tag{4.11a}$$

$$y(t, r) = \sum_{j=-\infty}^{\infty} H(r - j)x(t, j) + \sum_{j=-\infty}^{\infty} J(r - j)u(t, j). \tag{4.11b}$$

In (4.11a, b), $x(t, r) \in \mathbb{R}^n$ is the state at time t and spatial point $r \in \mathbb{Z}$, $u(t, r) \in \mathbb{R}^m$ is the input or control at time t and point r, and $y(t, r) \in \mathbb{R}^p$ is the output at time t and point r.

We will show that the state equation (4.11a) can be written as a differential equation in the Banach space $\ell^\infty(\mathbb{Z}, \mathbb{R}^n)$ with coefficients in the Banach algebra $\ell^1(\mathbb{Z}, \mathbb{R})$. First, for each fixed $t \in \mathbb{R}$, let u_t, x_t, y_t denote the vector-valued functions defined on $\mathbb{Z}$ given by

$$u_t(r) = u(t, r) \in \mathbb{R}^m, \qquad r \in \mathbb{Z},$$

$$x_t(r) = x(t, r) \in \mathbb{R}^n, \qquad r \in \mathbb{Z},$$

$$y_t(r) = y(t, r) \in \mathbb{R}^p, \qquad r \in \mathbb{Z}.$$

Now defining

$$\frac{dx_t}{dt}(r) = \frac{dx(t, r)}{dt}, \qquad r \in \mathbb{Z},$$

we can rewrite (4.11a) as a differential equation in $\ell^\infty(\mathbb{Z}, \mathbb{R}^n)$ given by

$$\frac{dx_t}{dt} = F *_r x_t + G *_r u_t, \qquad t > 0, \tag{4.12a}$$

with initial condition $x_0 \in \ell^\infty(\mathbb{Z}, \mathbb{R}^n)$, and where $*_r$ denotes convolution with respect to the spatial variable r. The output equation (4.11b) can also be rewritten in the form

$$y_t = H *_r x_t + J *_r u_t, \qquad t > 0. \tag{4.12b}$$

The system given by (4.12a, b) is clearly infinite dimensional since the state space $\ell^\infty(\mathbb{Z}, \mathbb{R}^n)$ is an infinite-dimensional Banach space. We shall study the solution to these equations in the next subsection.

A state representation $\Sigma = (F, G, H, J)$ also defines a m-input p-output linear spatially-distributed discrete-time system given by the state equations

$$x(k + 1, r) = \sum_{j = -\infty}^{\infty} F(r - j)x(k, j) + \sum_{j = -\infty}^{\infty} G(r - j)u(k, j) \tag{4.13a}$$

$$y(k, r) = \sum_{j = -\infty}^{\infty} H(r - j)x(k, j) + \sum_{j = -\infty}^{\infty} J(r - j)u(k, j). \tag{4.13b}$$

In Section 4.3 we show that state equations of the form (4.13a, b) arise by discretizing in time state equations of the form (4.11a, b).

As we did in the continuous-time case, we can rewrite the state equation (4.13a, b) in terms of the state space $\ell^\infty(\mathbb{Z}, \mathbb{R}^n)$: First, let u_k, x_k, y_k denote the vector-valued functions on $\mathbb{Z}$ defined by $u_k(r) = u(k, r)$,

$x_k(r) = x(k, r)$, and $y_k(r) = y(k, r)$. Then (4.13a, b) can be expressed in the form

$$x_{k+1} = F *_r x_k + G *_r u_k \tag{4.14a}$$

$$y_k = H *_r x_k + J *_r u_k, \tag{4.14b}$$

where again $*_r$ denotes convolution with respect to the spatial variable r.

4.2.2. *Solution of the State Equations*

Let $\Sigma = (F, G, H, J)$ be a state representation over $\ell^1(\mathbb{Z}, \mathbb{R})$. We shall first interpret Σ as the state model for a spatially-distributed continuous-time system given by the state equations (4.12a, b). To solve these state equations, we begin by considering the free (unforced) equation

$$\frac{dx_t}{dt} = F *_r x_t, \, t > 0, \, x_0 \in \ell^\infty(\mathbb{Z}, \mathbb{R}^n). \tag{4.15}$$

Let $B(\ell^\infty(\mathbb{Z}, \mathbb{R}^n))$ denote the Banach algebra consisting of all bounded linear maps from $\ell^\infty(\mathbb{Z}, \mathbb{R}^n)$ into itself. As noted in the previous subsection, the $n \times n$ matrix F defines an element of $B(\ell^\infty(\mathbb{Z}, \mathbb{R}^n))$ given by $v \to F * v$. We denote this operator by F; that is, we shall let F denote either the given $n \times n$ matrix over $\ell^1(\mathbb{Z}, \mathbb{R})$ or the element of $B(\ell^\infty(\mathbb{Z}, \mathbb{R}^n))$ defined by F.

Now for each fixed $t \geq 0$, let e^{Ft} denote the element of $B(\ell^\infty(\mathbb{Z}, \mathbb{R}^n))$ defined by

$$v \to e^{Ft} *_r v = v + \sum_{j=1}^{\infty} (F^j *_r v) \frac{t^j}{j!}, \tag{4.16}$$

where F^j is the j-fold convolution of F with itself. By a standard argument, one can show that the sequence of partial sums

$$\left\{ v + \sum_{j=1}^{N} (F^j *_r v) \frac{t^j}{j!} : N = 1, 2, \ldots \right\}$$

converges in $B(\ell^\infty(\mathbb{Z}, \mathbb{R}^n))$ to the infinite sum in (4.16), and that this convergence is uniform in $t \in [0, T]$, where T is any positive real number. Thus e^{Ft} is a well-defined element of $B(\ell^\infty(\mathbb{Z}, \mathbb{R}^n))$ for all $t \geq 0$. Note that we can also view e^{Ft} as an $n \times n$ matrix over $\ell^1(\mathbb{Z}, \mathbb{R})$ for each $t \geq 0$. In particular, the jth column of e^{Ft} is equal to $e^{Ft} *_r e_j$, where e_j is the jth unit vector of $\ell^\infty(\mathbb{Z}, \mathbb{R}^n)$.

Again from well-known results (see Dunford and Schwartz [4.7, pp.

614–615]), the set $\{e^{Ft}: t \geq 0\}$ with composition of operators is a uniformly continuous semigroup in $B(\ell^{\infty}(\mathbb{Z}, \mathbb{R}^n))$ with infinitesimal generator F. Further, the solution x_t to (4.15) with initial condition $x_0 \in \ell^{\infty}(\mathbb{Z}, \mathbb{R}^n)$ is given by

$$x_t = e^{Ft} *_r x_0, \qquad t \geq 0. \tag{4.17}$$

The proof that x_t given by (4.17) is the solution to (4.15) follows from the observation that

$$\frac{d}{dt} e^{Ft} = F *_r e^{Ft}.$$

We then have the following result on the complete solution to the state equation (4.12a).

THEOREM 4.1. *Suppose that the input u_t is an element of $\ell^{\infty}(\mathbb{Z}, \mathbb{R}^n)$ for all $t \geq 0$ and is locally integrable in the sense that*

$$\int_0^T \left[\sup_{r \in \mathbb{Z}} \|u(t, r)\| \right] dt < \infty \quad \text{for all} \quad T > 0.$$

Then if the initial state $x_0 \in \ell^{\infty}(\mathbb{Z}, \mathbb{R}^n)$, the state equation (4.12a) has a unique solution $x_t \in \ell^{\infty}(\mathbb{Z}, \mathbb{R}^n)$ for all $t > 0$ given by

$$x_t = e^{Ft} *_r x_0 + \int_0^t \left[e^{F(t-\tau)} *_r G *_r u_\tau \right] d\tau, \qquad t > 0, \tag{4.18}$$

where again $*_r$ denotes convolution with respect to the spatial variable r.

Proof: The theorem follows from a variation of parameters type argument: Assume there is a solution of the form

$$x_t = e^{Ft} *_r v_t, \qquad t > 0, \tag{4.19}$$

where $v_t \in \ell^{\infty}(\mathbb{Z}, \mathbb{R}^n)$ for all $t > 0$ and $v_0 = x_0$. Differentiating both sides of (4.19) with respect to t, we get

$$\frac{dx_t}{dt} = \left(\frac{d}{dt} e^{Ft} \right) *_r v_t + e^{Ft} *_r \frac{dv_t}{dt}$$

$$= F *_r x_t + e^{Ft} *_r \frac{dv_t}{dt}.$$

But

$$\frac{dx_t}{dt} = F *_r x_t + G *_r u_t,$$

so

$$e^{Ft} *_r \frac{dv_t}{dt} = G *_r u_t,$$

or

$$\frac{dv_t}{dt} = e^{-Ft} *_r G *_r u_t. \tag{4.20}$$

With the assumption on the input u_t, it follows that the solution v_t to (4.20) belongs to $\ell^\infty(\mathbb{Z}, \mathbb{R}^n)$ for all $t > 0$ and can be expressed in the form

$$v_t = v_0 + \int_0^t \left[e^{-F\tau} *_r G *_r u_\tau \right] d\tau, \, t > 0.$$

Then using (4.19) and the fact that the sequence of partial sums for e^{Ft} converges uniformly in $t \in [0, T]$ for any $T > 0$, we have that the solution x_t can be written in the form (4.18). Since $v_t \in \ell^\infty(\mathbb{Z}, \mathbb{R}^n)$ for all $t > 0$, it is clear from (4.19) that $x_t \in \ell^\infty(\mathbb{Z}, \mathbb{R}^n)$ for all $t > 0$. $\square$

By writing out the convolutions in (4.18), we can express the solution x_t in a more explicit form as follows. First, let $\Phi(t, r) = e^{Ft}(r)$, $r \in \mathbb{Z}$, where now we are viewing e^{Ft} as a $n \times n$ matrix over $\ell^1(\mathbb{Z}, \mathbb{R})$. Then evaluating both sides of (4.18) at the spatial point r, we have

$$x(t, r) = \sum_{j=-\infty}^{\infty} \Phi(t, r - j)x(0, j) + \tag{4.21}$$

$$\int_0^t \sum_{j=-\infty}^{\infty} \sum_{k=-\infty}^{\infty} \Phi(t - \tau, r - j)G(j - k)u(\tau, k) \, d\tau.$$

Now for each fixed $t \geq 0$, let Ω_t denote the $p \times m$ matrix over $\ell^1(\mathbb{Z}, \mathbb{R})$ defined by

$$\Omega_t = H *_r e^{Ft} *_r G.$$

Write $\Omega_t(r) = \Omega(t, r)$, $r \in \mathbb{Z}$. Then from (4.11b) and (4.21), if the initial state $x(0, r)$ is zero for all $r \in \mathbb{Z}$, the output response $y(t, r)$ resulting from input $u(t, r)$ can be expressed in the form

$$y(t, r) = \int_0^t \left[\sum_{j=-\infty}^{\infty} \Omega(t - \tau, r - j)u(\tau, j) \right] d\tau +$$

$$\sum_{j=-\infty}^{\infty} J(r - j)u(t, j). \tag{4.22}$$

From (4.22) we see that if $J = 0$, $\Omega(t, r)$ is the impulse response-function matrix of the system. Taking the joint Laplace/z transform of both sides of (4.22), we get the transfer-function representation

$$Y(s, z) = W(s, z)U(s, z),$$

where the transfer-function matrix $W(s, z)$ is given by

$$W(s, z) = \Omega(s, z) + J(z) = H(z)(sI - F(z))^{-1}G(z) + J(z). \tag{4.23}$$

From (4.23), we see that the transfer-function matrix $W(s, z)$ is a proper rational matrix function in s. If $J = 0$, $W(s, z)$ is a strictly proper rational matrix function in s. Note that $W(s, z)$ is not necessarily rational in z unless the z-transforms of the coefficient matrices F, G, H, J are rational in z, in which case $W(s, z)$ is rational in both s and z.

Let's now consider the solution to the state equations (4.14a, b) which arise in the discrete-time case. Unlike the continuous-time case, the solution to the discrete-time state equations can be derived very easily via iteration: If the input u_k belongs to $\ell^\infty(\mathbb{Z}, \mathbb{R}^m)$ for all $k \geq 0$ and the initial state x_0 belongs to $\ell^\infty(\mathbb{Z}, \mathbb{R}^n)$, the solution x_k to (4.14a) belongs to $\ell^\infty(\mathbb{Z}, \mathbb{R}^n)$ for all $k > 0$ and is given by

$$x_k = F^k *_r x_0 + \sum_{j=0}^{k-1} F^{k-j-1} *_r G *_r u_j, \qquad k = 1, 2, \ldots. \tag{4.24}$$

For $k = 0, 1, 2, \ldots$, let W_k denote the $p \times m$ matrix over $\ell^1(\mathbb{Z}, \mathbb{R})$ defined by

$$W_0 = J$$

$$W_k = H *_r F^{k-1} *_r G, \qquad k = 1, 2, \ldots.$$

Then the output response y_k resulting from input u_k with $x_0 = 0$ can be expressed in the form

$$y_k = \sum_{j=0}^{k} W_{k-j} *_r u_j, \qquad k = 0, 1, 2, \ldots. \tag{4.25}$$

Evaluating both sides of (4.25) at the spatial point $r \in \mathbb{Z}$, we have

$$y_k(r) = y(k, r) = \sum_{j=0}^{k} \sum_{q=-\infty}^{\infty} W(k - j, r - q)u(j, q), \quad (4.26)$$

where $W_k(r) = W(k, r)$. The matrix function $W(k, r)$ is the unit-pulse response function of the system.

Taking the two-dimensional z-transform of both sides of (4.26), we get the transfer-function representation

$$Y(z_1, z_2) = W(z_1, z_2)U(z_1, z_2),$$

where the transfer-function matrix $W(z_1, z_2)$ is given by

$$W(z_1, z_2) = H(z_2)(z_1 I - F(z_2))^{-1}G(z_2) + J(z_2). \quad (4.27)$$

4.2.3. *Input/output Description*

Consider a m-input p-output linear spatially-distributed continuous-time system with the input/output relationship

$$y(t, r) = \int_{-\infty}^{t} \left[\sum_{j=-\infty}^{\infty} W(t - \tau, r - j)u(\tau, j) \right] d\tau, \quad (4.28)$$

where $W(t, r)$ is the impulse response-function matrix of the system. For each $t > 0$, we assume that $W(t, \cdot)$ is a matrix over $\ell^1(\mathbb{Z}, \mathbb{R})$; that is, the entries of the impulse response function matrix are absolutely summable functions of the spatial variable r.

The transfer-function matrix $W(s, z)$ associated with the input/output relationship (4.28) is defined to be the joint Laplace/z transform of $W(t, r)$. We are not assuming that $W(s, z)$ is a rational function of s.

Now let A denote the set of all bi-infinite series in z with absolutely-summable real coefficients; that is,

$$\alpha(z) = \sum_{r=-\infty}^{\infty} \alpha_r z^{-r}, \qquad \alpha_r \in \mathbb{R} \quad \text{for all} \quad r,$$

is an element of A if and only if

$$\sum_{r=-\infty}^{\infty} |\alpha_r| < \infty.$$

With the usual operations, A is a commutative algebra with multiplicative identity. The elements of A can be viewed as the z-transforms of elements of $\ell^1(\mathbb{Z}, \mathbb{R})$. In fact, $\ell^1(\mathbb{Z}, \mathbb{R})$ and A are isomorphic as algebras.

Let $A[s]$ denote the ring of polynomials in s with coefficients in A. Then the transfer-function matrix $W(s, z)$ associated with the input/output relationship (4.28) is said to have a *right polynomial matrix factorization* if there exist a $p \times m$ matrix $P(s, z)$ over $A[s]$ and a $m \times m$ matrix $Q(s, z)$ over $A[s]$, with the leading coefficient of det $Q(s, z)$ invertible in A, such that

$$W(s, z) = P(s, z)Q(s, z)^{-1}.$$

There is an analogous definition for left polynomial matrix factorizations, but we will not consider this.

The system with transfer-function matrix $W(s, z)$ is said to be *realizable* if there exists a state representation $\Sigma = (F, G, H, J)$ over $\ell^1(\mathbb{Z}, \mathbb{R})$ such that

$$W(s, z) = H(z)(sI - F(z))^{-1}G(z) + J(z).$$

A realizability criterion is given in the following result.

PROPOSITION 4.1. *A system with transfer-function matrix $W(s, z)$ is realizable if and only if $W(s, z)$ has a right polynomial matrix factorization.*

Proof. The result is an immediate corollary of the realization theory of linear systems with coefficients in a commutative ring (e.g., see Sontag [4.8]).

EXAMPLE 4.2. Consider the seismic cable with transfer function $W(s, z)$ given by (4.9). Defining the state

$$x(t, r) = \begin{bmatrix} y(t, r) \\ \dfrac{dy(t, r)}{dt} \end{bmatrix},$$

where as before $y(t, r)$ is the position of the rth cable segment at time t, we have the obvious realization $\Sigma = (F, G, H, J)$, where

$$F(0) = \begin{bmatrix} 0 & 1 \\ -k/M & -c/M \end{bmatrix}, \; F(-1) = F(1) = \begin{bmatrix} 0 & 0 \\ k/2M & 0 \end{bmatrix},$$

$$F(r) = 0, \; |r| \geq 2,$$

$$G(0) = \begin{bmatrix} 0 \\ 1/M \end{bmatrix}, \quad G(r) = 0, \; |r| \geq 1.$$

$$H(0) = [1 \quad 0], \; H(r) = 0, \; |r| \geq 1, \; J(r) = 0 \quad \text{for all} \quad r.$$

The state equations associated with this realization are given by

$$\frac{dx(t, r)}{dt} = F(0)x(t, r) + F(1)[x(t, r - 1) + x(t, r + 1)] +$$

(4.29a)

$$G(0)u(t, r)$$

$$y(t, r) = H(0)x(t, r). \tag{4.29b}$$

The above definitions and result on realizability have obvious counter-parts in the discrete-time case which we will not consider here. We also wish to point out that we will be concerned with the existence of state representations possessing various properties that are useful in control. Before we get to this, we consider discretizations in time and then we study the notion of stability that will be used in the definition of stabiliz-ation by feedback.

4.3. DISCRETIZATIONS IN TIME

Given a spatially-distributed continuous-time system specified by the state equations (4.11a, b) or (4.12a, b), we now consider techniques for constructing a discretization in time. In addition to simplifying the com-putation of responses, a discretization in time is useful in the design of compensators that are discrete in both the time variable and the spatial variable. We shall first consider a discretization technique which is a natural generalization of the usual method for discretizing finite-dimen-sional continuous-time systems.

Let's begin with the state equation

$$\frac{dx_t}{dt} = F *_r x_t + G *_r u_t. \tag{4.30}$$

If the initial time is τ, the initial state $x_\tau \in \ell^\infty(\mathbb{Z}, \mathbb{R}^n)$, and the input u_t satisfies the hypothesis of Theorem 4.1, by a minor variation of Theorem

4.1 we have that the solution x_t to (4.30) can be expressed in the form

$$x_t = e^{F(t - \tau)} *_r x_\tau + \int_\tau^t e^{F(t - \lambda)} *_r G *_r u_\lambda \, d\lambda, \, t > \tau. \quad (4.31)$$

Now let T be a fixed positive real number and let $k \in \mathbb{Z}$. Then setting $t = kT + T$ and $\tau = kT$ in (4.30), we have the discretization

$$x_{kT + T} = e^{FT} *_r x_{kT} + \int_{kT}^{kT + T} e^{F(kT + T - \lambda)} *_r G *_r u_\lambda \, d\lambda. \quad (4.32)$$

If for each fixed $r \in \mathbb{Z}$, the input $u_t(r) = u(t, r)$ is approximately constant over each time interval $[kT, kT + T]$, we can write (4.32) in the form

$$x_{kT + T} = F_d *_r x_{kT} + G_d *_r u_{kT}, \quad (4.33a)$$

where

$$F_d = e^{FT} \quad \text{and} \quad G_d = \int_0^T e^{F\lambda} *_r G \, d\lambda.$$

As noted before, the matrix e^{FT} is over $\ell^1(\mathbb{Z}, \mathbb{R})$. Using the fact that the sequence of partial sums for e^{Ft} converges uniformly in $t \in [0, T]$, we have that the matrix G_d is also over $\ell^1(\mathbb{Z}, \mathbb{R})$. Finally, setting $t = kT$ in the output equation (4.12b), we have the discretized output equation

$$y_{kT} = H *_r x_{kT} + J *_r u_{kT}. \quad (4.33b)$$

The state representation $\Sigma_d = (F_d, G_d, H, J)$ with the associated state equations (4.33a, b) is a suitable discretization of the given continuous-time system. Note that since F_d and G_d are over $\ell^1(\mathbb{Z}, \mathbb{R})$, the discretization Σ_d is over $\ell^1(\mathbb{Z}, \mathbb{R})$.

The matrices F_d and G_d in the discretization Σ_d can be computed using the discrete Fourier transform as follows. Let N be a suitably large positive integer. We assume that $F(r)$ and $G(r)$ are both approximately zero for $|r| > N$. Let $\hat{F}_j$ and $\hat{G}_j$ denote the discrete Fourier transforms of F and G defined by

$$\hat{F}_j = \sum_{r = -N}^{N} F(r)e^{-ji\pi r/N}, \quad j = 0, \pm 1, \pm 2, \ldots, \pm N \quad (4.34a)$$

$$\hat{G}_j = \sum_{r = -N}^{N} G(r)e^{-ji\pi r/N}, \quad j = 0, \pm 1, \pm 2, \ldots, \pm N \quad (4.34b)$$

where $i = \sqrt{-1}$. The discrete Fourier transform of e^{FT} is equal to $e^{\hat{F}_j T}$.

Then $F_d = e^{FT}$ is given approximately by

$$F_d(r) = \frac{1}{2N+1} \sum_{j=-N}^{N} e^{\hat{F}_j T} e^{ji\pi r/N},$$

$$r = 0, \pm 1, \pm 2, \ldots, \pm N, \qquad (4.35)$$

and G_d is given approximately by

$$G_d(r) = \frac{1}{2N+1} \int_0^T \sum_{j=-N}^{N} \sum_{q=-N}^{N} e^{\hat{F}_j \lambda} \hat{G}_q e^{ij\pi r/N} \, d\lambda,$$

$$r = 0, \pm 1, \pm 2, \ldots, \pm N. \qquad (4.36)$$

The sums in (4.34a, b), (4.35), and (4.36) can be calculated using the FFT and the integral in (4.36) can be computed using numerical integration. Thus, we have an efficient method for computing the coefficient matrices F_d, G_d of the discretization Σ_d.

There is another discretization procedure which is based on writing the state equation (4.11a) in a block form: Given a fixed positive integer N, let $x_N(t, r)$ denote the nN-vector defined by

$$x_N(t, r) = \begin{bmatrix} x(t, rN) \\ x(t, rN + 1) \\ \vdots \\ x(t, rN + N - 1) \end{bmatrix}.$$

In terms of the nN-vector $x_N(t, r)$ we can rewrite the state equation (4.11a) in the form

$$\frac{dx_N(t, r)}{dt} = F_N(0)x_N(t, r) + \sum_{\substack{j=-\infty \\ j \neq r}}^{\infty} F_N(r - j)x_N(t, j) +$$

$$\sum_{j=-\infty}^{\infty} G_N(r - j)u(t, j). \qquad (4.37)$$

Here F_N is a $nN \times nN$ matrix over $\ell^1(\mathbb{Z}, \mathbb{R})$ and G_N is a $nN \times m$ matrix over $\ell^1(\mathbb{Z}, \mathbb{R})$. We can write the solution $x_N(t, r)$ to (4.37) in the form

$$x_N(t, r) = e^{F_N(0)(t - \tau)} x_N(\tau, r) +$$

$$\int_\tau^t e^{F_N(0)(t - \lambda)} \left[\sum_{\substack{j=-\infty \\ j \neq r}}^{\infty} F_N(r - j)x_N(\lambda, j) + \sum_{j=-\infty}^{\infty} G_N(r - j)u(\lambda, j) \right] d\lambda, \; t > \tau.$$

Taking $t = kT + T$, $\tau = kT$, and assuming that $x_N(t, j)$, $j \neq r$, and $u(t, r)$, all r, are approximately constant over the time interval $[kT + T, kT]$, we get the discretization

$$x_N(kT + T, r) = \sum_{j = -\infty}^{\infty} F_{N, d}(r - j)x_N(kT, j) +$$

$$\sum_{j = -\infty}^{\infty} G_{N, d}(r - j)u(kT, j), \qquad (4.38)$$

where

$$F_{N, d}(0) = e^{F_N(0)T}$$

$$F_{N, d}(r) = \int_0^T e^{F_N(0)\lambda} F_N(r) \, d\lambda, \qquad r \neq 0$$

$$G_{N, d}(r) = \int_0^T e^{F_N(0)\lambda} G_N(r) \, d\lambda, \qquad \text{all } r \in \mathbb{Z}.$$

The discrete-time approximation (4.38) is more accurate the larger N is. However, since the system matrices in (4.38) are of size $nN \times nN$, this discretization is not easy to compute if it is necessary to consider a value of N which is not small. In such cases we could use the first discretization method.

We shall apply the second discretization method to the seismic cable.

EXAMPLE 4.3. Consider the seismic cable which is given by the state equations (4.29a, b). Using the second discretization method with $N = 1$, we have that $F_N = F$ and $G_N = G$, and from (4.38), we get the discretization

$$x(kT + T, r) = F_{1, d}(0)x(kT, r) + F_{1, d}(1)[x(kT, r - 1) +$$

$$+ x(kT, r + 1)] + G_{1, d}(0)u(kT, r), \qquad (4.39)$$

where

$$F_{1, d}(0) = e^{F(0)T}$$

$$F_{1, d}(1) = \int_0^T e^{F(0)\lambda} F(1) \, d\lambda$$

$$G_{1, d}(0) = \int_0^T e^{F(0)\lambda} G(0) \, d\lambda.$$

The particular expressions for $F(0)$, $F(1)$, and $G(0)$ are given in Example 4.2. Later we shall use the discretization (4.39), so we need to compute the coefficient matrices $F_{1,d}(0)$, $F_{1,d}(1)$, $G_{1,d}(0)$. Using the standard Laplace-transform procedure, we get that

$$F_{1,d}(0) = e^{-cT/2M} \begin{bmatrix} \cos(aT) + (c/2aM)\sin(aT) & (1/a)\sin(aT) \\ -(k/aM)\sin(aT) & \cos(aT) - (c/2aM)\sin(aT) \end{bmatrix} \quad (4.40)$$

$$F_{1,d}(1) = \begin{bmatrix} e^{-cT/2M}[-(c/4aM)\sin(aT) - \tfrac{1}{2}\cos(aT)] + \tfrac{1}{2} & 0 \\ (k/2aM)e^{-cT/2M}\sin(aT) & 0 \end{bmatrix} \quad (4.41)$$

$$G_{1,d}(0) = (2/k) \text{ times the first column of } F_{1,d}(1). \quad (4.42)$$

Here

$$a = \sqrt{\frac{k}{M} - \frac{c^2}{4(M)^2}},$$

and we are assuming that a is real. As a check on the accuracy of this discretization, we could compute the discretization for $N = 2$ and compare the results. For $N = 2$, the system matrices in the block form are 4×4, so the computation of the discretization is best done on a computer. It would be interesting to compare these discretizations with that obtained by using (4.33a). We leave this to the interested reader.

4.4. REPRESENTATION IN TERMS OF A FAMILY OF FINITE-DIMENSIONAL SYSTEMS

Given a state representation $\Sigma = (F, G, H, J)$ over $\ell^1(\mathbb{Z}, \mathbb{R})$, it is very useful to consider the Fourier transform of the coefficient matrices F, G, H, J. In particular, we will show that by taking the Fourier transform of the system coefficients, we obtain a family of finite-dimensional systems parameterized by a real variable. We begin with a very brief description of the Fourier transform.

The Fourier transform of a function $v \in \ell^1(\mathbb{Z}, \mathbb{R})$ is a complex-valued continuous function $\hat{v}(\omega)$ defined on the interval $[0, 2\pi]$ given by

$$\hat{v}(\omega) = \sum_{r=-\infty}^{\infty} v(r) \, e^{-ir\omega}, \qquad \omega \in [0, 2\pi], \tag{4.43}$$

where again $i = \sqrt{-1}$. Since the function v is real valued, it follows from (4.43) that

$$\overline{\hat{v}(\omega)} = \hat{v}(2\pi - \omega) \quad \text{for all} \quad \omega \in [0, 2\pi],$$

where 'bar' denotes the complex conjugate. Conversely, let $\hat{\gamma}(\omega)$ be a continuous complex-valued function defined on $[0, 2\pi]$ such that $\overline{\hat{\gamma}(\omega)} = \hat{\gamma}(2\pi - \omega)$ for all $\omega \in [0, 2\pi]$. In general, the function $\hat{\gamma}$ does *not* have an inverse Fourier transform γ belonging to $\ell^1(\mathbb{Z}, \mathbb{R})$ (see Edwards [4.9]). However, if $\hat{\gamma}(\omega)$ is a continuously differentiable function of ω, $\hat{\gamma}$ does have an inverse Fourier transform γ belonging to $\ell^1(\mathbb{Z}, \mathbb{R})$. When it exists, the inverse Fourier transform γ of a continuous function $\hat{\gamma}$ is given by

$$\gamma(r) = (1/2\pi) \int_0^{2\pi} \hat{\gamma}(\omega) \, e^{ir\omega} \, d\omega, \qquad r \in \mathbb{Z}.$$

Now given a state representation $\Sigma = (F, G, H, J)$ over $\ell^1(\mathbb{Z}, \mathbb{R})$, we can apply the Fourier transform entry-by-entry to the matrices comprising Σ, which results in a family $\{\hat{\Sigma}(\omega) : \omega \in [0, 2\pi]\}$ of quadruples given by

$$\hat{\Sigma}(\omega) = (\hat{F}(\omega), \hat{G}(\omega), \hat{H}(\omega), \hat{J}(\omega)).$$

We can interpret the family $\{\hat{\Sigma}(\omega) : \omega \in [0, 2\pi]\}$ as either a family of finite-dimensional continuous-time systems or as a family of finite-dimensional discrete-time systems. In the continuous-time case, the family of systems is given by the collection of state equations

$$\frac{d\zeta(t, \omega)}{dt} = \hat{F}(\omega)\zeta(t, \omega) + \hat{G}(\omega)v(t, \omega) \tag{4.44a}$$

$$\gamma(t, \omega) = \hat{H}(\omega)\zeta(t, \omega) + \hat{J}(\omega)v(t, \omega), \; \omega \in [0, 2\pi], \tag{4.44b}$$

where $\zeta(t, \omega) \in \mathbb{C}^n$, $v(t, \omega) \in \mathbb{C}^m$, and $\gamma(t, \omega) \in \mathbb{C}^p$. Note that the systems defined by (4.44a, b) are over the field $\mathbb{C}$ of complex numbers; that is, in general the coefficients of these systems are complex numbers.

By the above constructions, we have that any state representation $\Sigma = (F, G, H, J)$ over $\ell^1(\mathbb{Z}, \mathbb{R})$ defines a family of finite-dimensional systems $\{\hat{\Sigma}(\omega):\omega \in [0, 2\pi]\}$ with coefficients in $\mathbb{C}$. We shall refer to $\{\hat{\Sigma}(\omega):\omega \in [0, 2\pi]\}$ as the *family of local systems* associated with Σ.

A key point is that we can study the system defined by a state representation Σ by studying the family of local systems defined by Σ. For example, we will show that if $\hat{\Sigma}(\omega)$ is stabilizable by state (or output) feedback for all $\omega \in [0, 2\pi]$, then the system with state representation Σ is stabilizable by state (or output) feedback with feedback compensators defined over $\ell^1(\mathbb{Z}, \mathbb{R})$. The difficult issue here is the "passage from the local to the global". In other words, if the localization of a property is true for each member of the family of local systems, we would like to know if the property is true for the given system. This question has been studied in various general frameworks with part of the emphasis on the existence of continuous canonical forms. For details on a portion of this past work, we refer the reader to Sontag [4.8], Hazewinkel [4.10], [4.11], [4.12], Byrnes [4.13], and Tannenbaum [4.14].

Now let $W(s, z)$ be a $p \times m$ transfer-function matrix associated with a system given by the input/output relationship (4.28). Suppose that the system is realizable so that the entries of $W(s, z)$ are rational functions in s with coefficients in the algebra A of absolutely summable bi-infinite series in z. Then we can evaluate $W(s, z)$ at $z = e^{i\omega}$, $\omega \in [0, 2\pi]$, and this evaluation defines a family $\{W(s, \omega):\omega \in [0, 2\pi]\}$ of transfer-function matrices whose entries are ratios of polynomials belonging to $\mathbb{C}[s]$, the ring of polynomials in s with complex coefficients. We can then study $W(s, z)$ in terms of the family $\{W(s, \omega):\omega \in [0, 2\pi]\}$. Again we can ask when local properties are equivalent to global properties.

4.5. STABILITY

In this section we first study stability in terms of a state representation. Then we consider input/output stability.

4.5.1. *Internal Stability*

Again let $\Sigma = (F, G, H, J)$ be a state representation over $\ell^1(\mathbb{Z}, \mathbb{R})$. Recall that when Σ is interpreted as the state model of a spatially-distributed continuous-time system, the free behavior of the system is given by

$$\frac{dx_t}{dt} = F *_r x_t, \, t > 0, \tag{4.45}$$

with initial state $x_0 \in \ell^\infty(\mathbb{Z}, \mathbb{R}^n)$. When Σ is interpreted as the state model of a spatially-distributed discrete-time system, the free behavior is given by

$$x_{k+1} = F *_r x_k, \qquad k = 0, 1, 2, \ldots \tag{4.46}$$

with initial state $x_0 \in \ell^\infty(\mathbb{Z}, \mathbb{R}^n)$.

DEFINITION 4.2. The system with state representation $\Sigma = (F, G, H, J)$, or the matrix F, is said to be *c-stable* if

$$\sup_{r \in \mathbb{Z}} \|x(t, r)\| \to 0 \quad \text{as} \quad t \to \infty \quad \text{for all} \quad x_0 \in \ell^\infty(\mathbb{Z}, \mathbb{R}^n),$$

where $x(t, r) = x_t(r)$ is the solution to (4.45) with initial state x_0. The system with state representation Σ, or the matrix F, is said to be *d-stable* if

$$\sup_{r \in \mathbb{Z}} \|x(k, r)\| \to 0 \quad \text{as} \quad k \to \infty \ (k = 1, 2, \ldots)$$

$$\text{for all} \quad x_0 \in \ell^\infty(\mathbb{Z}, \mathbb{R}^n),$$

where $x(k, r) = x_k(r)$ is the solution to (4.46) with initial state x_0.

Here the prefix 'c' (resp. 'd') stands for continuous time (resp., discrete time). It is important to note that by Definition 4.2, c-stability or d-stability implies that the state $x(t, r)$ (or $x(k, r)$) converges to zero *uniformly* in the spatial variable r. More precisely, c-stability implies that for any $\epsilon > 0$, there is a t_ϵ such that

$$\|x(t, r)\| \leq \epsilon \quad \text{for all} \quad t > t_\epsilon \quad \text{and all} \quad r \in \mathbb{Z},$$

where $x(t, r)$ is the state at time t and spatial point r resulting from an initial state $x(0, r)$ which is a bounded function of r. Convergence to zero uniformly in the spatial variable r is a very desirable property. In particular, if we only had convergence to zero pointwise in r, we could not guarantee that the state $x(t, r)$ is small across the entire distributed structure for a suitably large value of t.

We shall first characterize stability in terms of two different matrix norms defined as follows. Given a $n \times n$ matrix P over $\ell^1(\mathbb{Z}, \mathbb{R})$, let $\|P\|_1$ and $\|P\|_\infty$ denote the matrix norms defined by

$$\|P\|_1 = \sup\{\|P * x\|_1 : x \in \ell^1(\mathbb{Z}, \mathbb{R}^n), \|x\|_1 = 1\}. \tag{4.47}$$

$$\|P\|_\infty = \sup\{\|P * x\|_\infty : x \in \ell^\infty(\mathbb{Z}, \mathbb{R}^n), \|x\|_\infty = 1\}. \qquad (4.48)$$

It can be shown that $\|P * x\|_\infty \le \|P\|_1 \|x\|_\infty$ for all $x \in \ell^\infty(\mathbb{Z}, \mathbb{R}^n)$, and thus

$$\|P\|_\infty \le \|P\|_1 \text{ for any } P \text{ over } \ell^1(\mathbb{Z}, \mathbb{R}).$$

PROPOSITION 4.2. *For the system with state representation* Σ, *the following are equivalent.*
   (i)   *The system is c-stable*;
   (ii) $\|e^{Ft}\|_\infty \to 0$ *as* $t \to \infty$;
   (iii) $\|e^{Ft}\|_1 \to 0$ *as* $t \to \infty$.

Proof. By the results in Section 4.2.2, $x_t = e^{Ft} *_r x_0$ is the solution to (4.45) with initial condition $x_0 \in \ell^\infty(\mathbb{Z}, \mathbb{R}^n)$. Letting x_0 range over the unit vectors of $\ell^\infty(\mathbb{Z}, \mathbb{R}^n)$, we have that c-stability is equivalent to condition (ii). The implication (iii) $\Rightarrow$ (i) follows from the relationship $\|e^{Ft} *_r x_0\| \le \|e^{Ft}\|_1 \|x_0\|_\infty$. The proof that (i) $\Rightarrow$ (iii) is similar to a proof in Kamen and Green [4.15, p. 600]: The condition $\|e^{Ft} *_r x_0\|_\infty \to 0$ for all $x_0 \in \ell^\infty(\mathbb{Z}, \mathbb{R}^n)$ implies that $e^{Ft} \to 0$ weakly in the Banach algebra of $n \times n$ matrices over $\ell^1(\mathbb{Z}, \mathbb{R})$ with the norm (4.47). But since $\ell^1(\mathbb{Z}, \mathbb{R})$ has the Schur property, $e^{Ft} \to 0$ in the norm (4.47); that is, weak convergence implies norm convergence in this particular Banach algebra. $\square$

The corresponding result in the discrete-time case is given below. The proof is similar and is thus omitted.

PROPOSITION 4.3. For the system with state representation $\Sigma = (F, G, H, J)$, the following are equivalent:
   (i)   The system is d-stable;
   (ii) $\|F^k\|_\infty \to 0$ as $k \to \infty$ $(k = 1, 2, \ldots)$;
   (iii) $\|F^k\|_1 \to 0$ as $k \to \infty$.

Using Propositions 4.2 and 4.3, we have the following result.

PROPOSITION 4.4. *The matrix F is c-stable if and only if* e^F *is d-stable.*

Proof. Clearly, $\|e^{Ft}\|_1 \to 0$ implies that $\|(e^F)^k\|_1 \to 0$, and thus c-stability of F implies d-stability of e^F. Conversely, suppose that $\|(e^F)^k\|_1 \to 0$ as $k \to \infty$ $(k = 1, 2, \ldots)$. Then there exists a positive integer q such that $\|e^{Fq}\|_1 < 1$. Thus, $\|(e^{Fq})^t\|_1 \to 0$ as $t \to \infty$, which implies that $\|e^{Ft}\|_1 \to 0$ as $t \to \infty$. $\square$

We shall now give local criteria for stability. First, given a $n \times n$ matrix P over $\ell^1(\mathbb{Z}, \mathbb{R})$ with Fourier transform $\hat{P}(\omega)$, for each $\omega \in [0, 2\pi]$ let $\lambda_j(\hat{P}(\omega))$, $j = 1, 2, \ldots, n$ denote the eigenvalues of $\hat{P}(\omega)$.

THEOREM 4.2. *The system with state representation* $\Sigma = (F, G, H, J)$ *is d-stable if and only if*

$$|\lambda_j(\hat{F}(\omega))| < 1, \qquad j = 1, 2, \ldots, n, \qquad \omega \in [0, 2\pi]. \quad (4.49)$$

The system is c-stable if and only if

$$\text{Re } \lambda_j(\hat{F}(\omega)) < 0, \qquad j = 1, 2, \ldots, n, \qquad \omega \in [0, 2\pi]. \quad (4.50)$$

Proof. By the results of Kamen and Green [4.15, pp. 587–589], (4.49) is equivalent to $\|F^k\|_1 \to 0$. Thus by Proposition 4.3, d-stability is equivalent to (4.49). To prove the second part of the theorem, first note that by Proposition 4.4, c-stability of F is equivalent to d-stability of e^F. Now the Fourier transform of e^F is equal to $e^{\hat{F}}$, and thus by the first part of the theorem, d-stability of e^F is equivalent to

$$|\lambda_j(e^{\hat{F}(\omega)})| < 1, \qquad j = 1, 2, \ldots, n, \qquad \omega \in [0, 2\pi]. \quad (4.51)$$

By a standard argument, (4.51) is equivalent to (4.50). $\square$

By Theorem 4.2, we see that c-stability or d-stability of the system defined by Σ is equivalent to stability of each system in the family of finite-dimensional local systems $\{\hat{\Sigma}(\omega): \omega \in [0, 2\pi]\}$. Hence, stability can be checked by applying existing stability tests for finite-dimensional systems (with coefficients in $\mathbb{C}$).

EXAMPLE 4.4. Again consider the seismic cable with system matrix F, where

$$F(0) = \begin{bmatrix} 0 & 1 \\ -k/M & -c/M \end{bmatrix}, \, F(-1) = F(1) = \begin{bmatrix} 0 & 0 \\ k/2M & 0 \end{bmatrix},$$

$$F(r) = 0, \, |r| \geq 2.$$

The Fourier transform $\hat{F}(\omega)$ is given by

$$\hat{F}(\omega) = \begin{bmatrix} 0 & 1 \\ -(k/M)(1 - \cos \omega) & -c/M \end{bmatrix}.$$

When $\omega = 0$, the eigenvalues of $\hat{F}(\omega)$ are 0 and $-c/M$, and thus by Theorem 4.2, the cable is not c-stable.

4.5.2. *Input/Output Stability*

Consider the spatially-distributed continuous-time system with input/
output relationship

$$y(t, r) = \int_{-\infty}^{t} \left[\sum_{j=-\infty}^{\infty} W(t - \tau, r - j)u(\tau, j) \right] d\tau. \qquad (4.52)$$

As in Section 4.2.3, we assume that the impulse-response function matrix
$W(t, \cdot)$ is over $\ell^1(\mathbb{Z}, \mathbb{R})$ for all $t > 0$.

DEFINITION 4.3. The system with input/output relationship (4.52) is
bounded-input bounded-output (BIBO) stable if

$$\sup_{r \in \mathbb{Z}} \|y(t, r)\| \le c_y < \infty \quad \text{for all} \quad t,$$

whenever

$$\sup_{r \in \mathbb{Z}} \|u(t, r)\| \le c_u < \infty \quad \text{for all} \quad t.$$

We shall give a sufficient local criterion for BIBO stability. Let $W(s, z)$
denote the transfer-function matrix associated with the system defined by
(4.52).

THEOREM 4.3. *Suppose that the transfer-function matrix $W(s, z)$ has the
right polynomial matrix factorization*

$$W(s, z) = P(s, z)Q(s, z)^{-1},$$

with

$$\det Q(s, e^{i\omega}) \ne 0, \qquad \text{Re } s \ge 0, \qquad \omega \in [0, 2\pi]. \qquad (4.53)$$

Then the system with transfer-function matrix $W(s, z)$ is BIBO stable.

Proof. Combining Proposition 4.1, Theorem 4.2, and results on the
realization of systems over rings, we have that the existence of a right
polynomial factorization with the property (4.53) implies that there is a
c-stable realization with state representation $\Sigma = (F, G, H, J)$ over
$\ell^1(\mathbb{Z}, \mathbb{R})$. We shall assume that $J = 0$. (The direct-feed coefficient matrix
J has no effect on BIBO stability.) From the results in Section 4.2.2, the
impulse-response function matrix $W(t, r) = W_t(r)$ is given by $W_t =$

$H *_r e^{Ft} *_r G, t > 0$. Since the realization is c-stable, $\|e^{Ft}\|_1 \to 0$. It follows that there exists a positive constant $b < \infty$ such that

$$\int_0^t \|H *_r e^{F\lambda} *_r G\|_1 \, d\lambda < b \quad for \; all \quad t > 0. \tag{4.54}$$

Now we can write (4.52) in the form (assuming $u_\tau = 0$ for all $\tau < 0$)

$$y_t = \int_0^t [W_{t-\tau} *_r u_\tau] \, d\tau. \tag{4.55}$$

Taking the norm of both sides of (4.55), we have that

$$\|y_t\|_\infty \leq \left[\int_0^t \|W_{t-\tau}\|_1 \, d\tau \right] \left[\sup_{\tau \in [0, \, t]} \|u_\tau\|_\infty \right], \quad all \quad t > 0.$$

Then using (4.54), we get that

$$\|y_t\|_\infty \leq b \left[\sup_{\tau \in [0, \, t]} \|u_\tau\|_\infty \right], \quad t > 0.$$

Hence, bounded inputs produce bounded outputs, and thus the system is BIBO stable. $\square$

It is interesting to note that the criterion for BIBO stability in Theorem 4.3 is *not* necessary in general. In other words, there are BIBO stable systems whose transfer-function matrix $W(s, z)$ does not have a right polynomial matrix factorization with the property (4.53). This is equivalent to saying that there are BIBO stable systems which do not have a c-stable realization. Thus, our notion of internal stability is in general stronger than BIBO stability. This situation is similar to the well-known fact in the theory of two-dimensional digital filters that the stability criterion $Q(z_1, z_2) \neq 0$, $|z_1| \leq 1$ and $|z_2| \leq 1$, is in general stronger than BIBO stability (see Goodman [4.16]). The underlying issue here is the possibility of nonessential singularities of the second kind on the boundary of the stability region in the transfer-function matrix $W(s, z)$. We shall not pursue this here.

4.6. REACHABILITY AND STABILIZABILITY

In this section we define the notions of reachability and stabilizability for spatially-distributed systems given by a state representation Σ over $\ell^1(\mathbb{Z}, \mathbb{R})$. It is shown that reachability implies stabilizability by state feedback, and that for systems with coefficients in certain subrings of

$\ell^1(\mathbb{Z}, \mathbb{R})$, reachability implies pole assignability. We begin with the definition of a *-operation on $\ell^1(\mathbb{Z}, \mathbb{R})$.

Given $v \in \ell^1(\mathbb{Z}, \mathbb{R})$, let v^* denote the element of $\ell^1(\mathbb{Z}, \mathbb{R})$ defined by $v^*(r) = v(-r)$, $r \in \mathbb{Z}$. Letting $\hat{v}$ denote the Fourier transform of v, we have that

$$\widehat{v^*}(\omega) = \overline{\hat{v}(\omega)}, \qquad \omega \in [0, 2\pi], \tag{4.56}$$

where as before, the bar denotes the complex conjugate. Now let P be a $n \times n$ matrix over $\ell^1(\mathbb{Z}, \mathbb{R})$. Letting p_{ij} denote the ij-entry of P, we shall write $P = (p_{ij})$. The *-operation on $\ell^1(\mathbb{Z}, \mathbb{R})$ can be extended to matrices over $\ell^1(\mathbb{Z}, \mathbb{R})$ by defining

$$P^* = (p_{ji}{}^*). \tag{4.57}$$

Using (4.56), we have

$$\widehat{P^*}(\omega) = \hat{P}(\omega)^*, \qquad \omega \in [0, 2\pi], \tag{4.58}$$

where $\hat{P}(\omega)^*$ is the conjugate transpose of $\hat{P}(\omega)$. (This dual use of the * symbol should not be confusing.)

Now consider a system with state representation $\Sigma = (F, G, H, J)$ over $\ell^1(\mathbb{Z}, \mathbb{R})$. To simplify the terminology, we shall refer to Σ as the system, although Σ is really the state representation of a system.

Given a positive integer q, the *q-step reachability matrix* R_q of the system $\Sigma = (F, G, H, J)$ is the $n \times qn$ matrix over $\ell^1(\mathbb{Z}, \mathbb{R})$ defined by

$$R_q = [G \; FG \; \ldots \; F^{q-1}G].$$

Here and in the following development, we will often omit the symbol $*$ for convolution (e.g., $FG = F_*G$ in the above expression for R_q).

DEFINITION 4.4. The system $\Sigma = (F, G, H, J)$ is *reachable* if the n-step reachability matrix R_n has a right inverse over $\ell^1(\mathbb{Z}, \mathbb{R})$; that is, there exists a $nm \times n$ matrix S over $\ell^1(\mathbb{Z}, \mathbb{R})$ such that $R_n S = I$. The system is *locally reachable* if rank $\hat{R}_n(\omega) = n$ for all $\omega \in [0, 2\pi]$, where $\hat{R}_n$ is the Fourier transform of R_n.

Our notion of reachability is sometimes referred to as 'ring reachability' in the theory of linear systems whose coefficients belong to a commutative ring (see Sontag [4.8]). Local reachability is equivalent to requiring that each system in the family of finite-dimensional local systems $\{\hat{\Sigma}(\omega): \omega \in [0, 2\pi]\}$ is reachable in the usual sense.

THEOREM 4.4. *For the system Σ, the following are equivalent:*
  (i) Σ *is locally reachable*;
  (ii) Σ *is reachable*;
  (iii) *the $n \times n$ matrix $R_q(R_q^*)$ has an inverse over $\ell^1(\mathbb{Z}, \mathbb{R})$ for all $q \geq n$.*

Proof. If $R_n(R_n^*)$ is invertible over $\ell^1(\mathbb{Z}, \mathbb{R})$, then $R_n^*(R_n(R_n^*))^{-1}$ is a right inverse over $\ell^1(\mathbb{Z}, \mathbb{R})$ for R_n, and thus (iii) $\Rightarrow$ (ii). Suppose that R_n has a right inverse S over $\ell^1(\mathbb{Z}, \mathbb{R})$. Then $\hat{R}_n(\omega)\hat{S}(\omega) = I$ for all $\omega \in [0, 2\pi]$, which implies that rank $\hat{R}_n(\omega) = n$ for all ω. Hence, (ii) $\Rightarrow$ (i). Finally, suppose that rank $\hat{R}_n(\omega) = n$ for all $\omega \in [0, 2\pi]$. Then $\hat{R}_q(\omega)\hat{R}_q(\omega)^*$ is invertible for all $\omega \in [0, 2\pi]$ and all $q \geq n$. By (4.58), $\hat{R}_q(\omega)^* = \widehat{R_q^*}(\omega)$, and thus $\hat{R}_q(\omega)\widehat{R_q^*}(\omega)$ is invertible for all ω and all $q \geq n$. By Wiener's invertibility theorem (see Berberian [4.17, p. 267]), $R_q(R_q^*)$ has an inverse over $\ell^1(\mathbb{Z}, \mathbb{R})$ for all $q \geq n$, so (i) $\Rightarrow$ (iii). $\square$

We should note that the equivalence between reachability and local reachability follows directly from a result of Sontag [4.8, p. 19]. Here one must use the fact that the maximal ideals of $\ell^1(\mathbb{Z}, \mathbb{C})$ can be put into a one-to-one correspondence with the points of $[0, 2\pi]$.

There is another characterization of reachability which is useful in the case where Σ is interpreted as a continuous-time system: Given a fixed positive real number T, let X_T denote the *T-second reachability Grammian* defined by

$$X_T = \int_0^T \left[e^{-Ft} *_r G *_r G^* *_r e^{-F^* t} \right] dt.$$

Again using the fact that the sequence of partial sums for e^{-Ft} and $e^{-F^* t}$ converges uniformly in $t \in [0, T]$, we can show that X_T is a $n \times n$ matrix over $\ell^1(\mathbb{Z}, \mathbb{R})$.

PROPOSITION 4.5. *The system Σ is reachable if and only if X_T has an inverse over $\ell^1(\mathbb{Z}, \mathbb{R})$.*

Proof. From known results on finite-dimensional continuous-time systems, Σ is locally reachable if and only if $\hat{X}_T(\omega)$ is invertible for all $\omega \in [0, 2\pi]$. Again by Wiener's invertibility theorem, we have that Σ is locally reachable if and only if X_T is invertible over $\ell^1(\mathbb{Z}, \mathbb{R})$. But local reachability is equivalent to reachability. $\square$

We now consider the notion of stabilizability.

DEFINITION 4.5. The system $\Sigma = (F, G, H, J)$ is *c-stabilizable* (resp. *d-stabilizable*) if there exists a $m \times n$ matrix K over $\ell^1(\mathbb{Z}, \mathbb{R})$ such that

$F - GK$ is c-stable (d-stable). The system is *locally c-stabilizable* (*locally d-stabilizable*) if for each fixed $\omega \in [0, 2\pi]$, there exists a $m \times n$ matrix K_ω over $\mathbb{C}$ such that $\hat{F}(\omega) - \hat{G}(\omega)K_\omega$ has all eigenvalues in the open left-half plane (all eigenvalues in the open unit disc).

The property of c-stabilizability implies that there is a distributed state feedback control law

$$u(t, r) = - \sum_{j=-\infty}^{\infty} K(r - j)x(t, j),$$

such that the resulting closed-loop system is c-stable. Local c-stabilizability is equivalent to requiring that each system in the family of local systems $\{\hat{\Sigma}(\omega): \omega \in [0, 2\pi]\}$ can be stabilized using state feedback over $\mathbb{C}$.

Using Hautus' results (see [4.18]) on the stabilizability of finite-dimensional systems, we have the following necessary and sufficient condition for local stabilizability.

PROPOSITION 4.6. The system $\Sigma = (F, G, H, J)$ is locally c-stabilizable if and only if

$$\text{rank}[sI - \hat{F}(\omega) \quad \hat{G}(\omega)] = n, \quad \text{Re } s \geq 0, \quad \omega \in [0, 2\pi].$$

The system Σ is locally d-stabilizable if and only if

$$\text{rank}[zI - \hat{F}(\omega) \quad \hat{G}(\omega)] = n, \quad |z| \geq 1, \quad \omega \in [0, 2\pi].$$

It is easy to see that stabilizability implies local stabilizability. However, whether or not the converse is true is a nontrivial question. We answer this in the next section. Here we shall consider a sufficient condition for stabilizability.

THEOREM 4.5. *Suppose Σ is reachable. Then Σ is both c-stabilizable and d-stabilizable. In particular, with*

$$K_c = G^*(X_T^{-1}), \tag{4.59}$$

the matrix $F - GK_c$ is c-stable, and with

$$K_d = G^*(F^*)^n(R_{n+1}(R_{n+1}^*))^{-1}(F^{n+1}), \tag{4.60}$$

the matrix $F - GK_d$ is d-stable. (The convolution symbols are omitted in (4.59) and (4.60)).

Proof. If Σ is reachable, by Theorem 4.4 and Proposition 4.5, both X_T and $R_{n+1}(R_{n+1}^*)$ are invertible over $\ell^1(\mathbb{Z}, \mathbb{R})$. Thus K_c and K_d given by

(4.59) and (4.60) are well defined and their entires belong to $\ell^1(\mathbb{Z}, \mathbb{R})$. By the results of Kleinman [4.19] on the stabilizability of finite-dimensional continuous-time systems, $\hat{K}_c(\omega)$ is a stabilizing feedback matrix for $\hat{\Sigma}(\omega)$; i.e., the eigenvalues of $\hat{F}(\omega) - \hat{G}(\omega)\hat{K}_c(\omega)$ are in the open left-half plane. Thus by Theorem 4.2, $F - GK_c$ is c-stable. By the results of Kamen and Khargonekar [4.6, Theorem 4.6], for each $\omega \in [0, 2\pi]$, $\hat{K}_d(\omega)$ is a stabilizing feedback; i.e., the eigenvalues of $\hat{F}(\omega) - \hat{G}(\omega)\hat{K}_d(\omega)$ are in the open unit disc. Hence by Theorem 4.2, $F - GK_d$ is d-stable. $\square$

We should point out that the result that reachability implies c-stabilizability also follows by application of the results of Byrnes [4.20], [4.21] to systems with coefficients in the algebra $\ell^1(\mathbb{Z}, \mathbb{R})$ (see also Delchamps [4.22]).

As is the case for finite-dimensional systems, reachability of a system Σ over $\ell^1(\mathbb{Z}, \mathbb{R})$ is in general a much stronger property than stabilizability. In fact, for systems with coefficients in certain subrings of $\ell^1(\mathbb{Z}, \mathbb{R})$, reachability implies *pole assignability*:

THEOREM 4.6. (*Application of Morse's theorem*): *Let Λ by a subring of $\ell^1(\mathbb{Z}, \mathbb{R})$ which is a principal ideal domain (PID). Suppose that Σ is a reachable system over Λ. Then given any finite set $\{\alpha_1, \alpha_2, \ldots, \alpha_n\}$ of elements of Λ, there is a $m \times n$ matrix K over Λ such that*

$$\det(sI - F + GK) = (s - \alpha_1)(s - \alpha_2) \ldots (s - \alpha_n).$$

Proof. The theorem is an immediate consequence of Morse's theorem [4.23] for systems over the polynomial ring $\mathbb{R}[z]$, plus Sontag's observation [4.8, p. 20] that Morse's theorem holds for reachable systems over any PID. $\square$

There are two interesting subrings of $\ell^1(\mathbb{Z}, \mathbb{R})$ which are PID's. One of them is the subring consisting of all elements of $\ell^1(\mathbb{Z}, \mathbb{R})$ with finite support. This subring is isomorphic to the ring $\mathbb{R}[z, z^{-1}]$ of polynomials in z and z^{-1} with coefficients in the reals $\mathbb{R}$. The second subring which is a PID is the subring of $\ell^1(\mathbb{Z}, \mathbb{R})$ which is isomorphic to the ring $\mathbb{R}_{\bar{T}}(z)$ of rational functions in z with real coefficients and with no poles on the unit circle $\bar{T}$. (This case was pointed out to the writer by P. P. Khargonekar.) Given such a rational function $a(z)/b(z)$, the corresponding element of $\ell^1(\mathbb{Z}, \mathbb{R})$ is defined as follows. First, factor $b(z)$ into the product $b_1(z)b_2(z)$, where the zeros of $b_1(z)$ are in the open unit disc and the zeros of $b_2(z)$ are outside the closed unit disc. Let δ_1 denote the unique element of $\ell^1(\mathbb{Z}, \mathbb{R})$ with support bounded on the left and with two-sided z-

transform equal to $a(z)/b_1(z)$. Let δ_2 denote the unique element of $\ell^1(\mathbb{Z}, \mathbb{R})$ with support bounded on the right and with two-sided z-transform equal to $1/b_2(z)$. Then $\delta_1 * \delta_2$ is the element of $\ell^1(\mathbb{Z}, \mathbb{R})$ which corresponds to the rational function $a(z)/b(z)$.

By Theorem 4.6, for reachable systems $\Sigma = (F, G, H, J)$, with the z-transforms of the entries of F and G belonging to $\mathbb{R}[z, z^{-1}]$ or $\mathbb{R}_{\tilde{\mathcal{P}}}(z)$, it is possible to place the poles in the sense defined in Theorem 4.6. In fact, by choosing the α_i in Theorem 4.6 to be real numbers, the characteristic polynomial $\det(sI - F + GK)$ is a polynomial in s with real coefficients, so in a sense the closed-loop system is finite dimensional (that is, we can remove all spatial dependency from the characteristic polynomial). We should also mention that Morse's proof [4.23] is constructive, so that for systems with transformed coefficients in $\mathbb{R}[z, z^{-1}]$ or $\mathbb{R}_{\tilde{\mathcal{P}}}(z)$, we can actually compute a feedback matrix K which places the closed-loop poles.

Finally, for single-input systems with coefficients in any commutative ring, it is known (Sontag [4.8]) that reachability implies pole assignability (and 'coefficient assignability'). Hence, any single-input $(m = 1)$ reachable system Σ over $\ell^1(\mathbb{Z}, \mathbb{R})$ is pole assignable.

EXAMPLE 4.5. Consider the seismic cable with two-step reachability matrix R_2 whose Fourier transform $\hat{R}_2$ is given by

$$\hat{R}_2(\omega) = [\hat{G}(\omega) \quad \hat{F}(\omega)\hat{G}(\omega)] = \begin{bmatrix} 0 & 1/M \\ 1/M & -c/M \end{bmatrix}$$

Clearly, rank $\hat{R}_2(\omega) = 2$ for all $\omega \in [0, 2\pi]$, and thus the cable is a reachable system. Since there is only one input, the cable system is pole assignable.

4.7. THE RICCATI EQUATION AND STABILIZABILITY

In this section we study the problem of stabilizability in terms of Riccati equations associated with the given system.

Let $\Sigma = (F, G, H, J)$ be a system over $\ell^1(\mathbb{Z}, \mathbb{R})$ with the associated family of local systems $\{\hat{\Sigma}(\omega): \omega \in [0, 2\pi]\}$, where $\hat{\Sigma}(\omega) = (\hat{F}(\omega), \hat{G}(\omega), \hat{H}(\omega), \hat{J}(\omega))$. Let V be a $n \times n$ matrix over $\ell^1(\mathbb{Z}, \mathbb{R})$ such that $\hat{V}(\omega)$ is a positive definite Hermitian matrix for every $\omega \in [0, 2\pi]$. We really only need that $\hat{V}(\omega)$ is positive semidefinite Hermitian for all ω (plus a detectability condition), but we will not consider this here.

Now consider the following two families of algebraic Riccati equations (ARE's):

$$P_\omega \hat{F}(\omega) + \hat{F}(\omega)^* P_\omega - P_\omega \hat{G}(\omega)(\hat{G}(\omega)^*)P_\omega + \hat{V}(\omega) = 0,$$
$$\omega \in [0, 2\pi] \quad (4.61)$$

$$\hat{F}(\omega)^* P_\omega \hat{F}(\omega) - P_\omega - \hat{F}(\omega)^* P_\omega \hat{G}(\omega)[\hat{G}(\omega)^* P_\omega \hat{G}(\omega) +$$
$$+ I]^{-1} \hat{G}(\omega)^* P_\omega \hat{F}(\omega) + \hat{V}(\omega) = 0$$
$$\omega \in [0, 2\pi]. \quad (4.62)$$

The ARE (4.61) arises in the continuous-time interpretation of Σ and the ARE (4.62) arises in the discrete-time interpretation of Σ. From well-known results on the Riccati-equation theory of finite-dimensional systems (e.g., see Kwakernaak and Sivan [4.24]), we have the following result.

PROPOSITION 4.7. The system Σ is locally c-stabilizable (resp. locally d-stabilizable) if and only if the ARE (4.61) (the ARE (4.62)) has a unique positive definite Hermitian solution P_ω for all $\omega \in [0, 2\pi]$.

Now consider the (global) Riccati difference equation (RDE)

$$P_{k+1} = F^* P_k F - F^* P_k G[G^* P_k G + I]^{-1} G^* P_k F + V,$$
$$k = 0, 1, 2, \ldots, \quad (4.63)$$

with the initial condition $P_0 = I$. It is easy to see that the solution sequence $\{P_k\}$ to (4.63) is a sequence of $n \times n$ matrices over $\ell^1(\mathbb{Z}, \mathbb{R})$. In terms of a finite-step solution to (4.63), we have the following result on stabilizability.

THEOREM 4.7. *Suppose that Σ is locally d-stabilizable. Then there exists a positive integer q such that with*

$$K_q = (G^* P_q G + I)^{-1} G^* P_q F, \quad (4.64)$$

the matrix $F - GK_q$ is d-stable.

Proof. The theorem follows directly from the results of Green and Kamen [4.4] or the results of Kamen and Khargonekar [4.6]. As both proofs are nontrivial, we refer the interested reader to either of these papers for the details. □

Combining Theorem 4.7 and Proposition 4.6, we have the following central result.

THEOREM 4.8 (*Green-Kamen-Khargonekar* [*4.4*], [*4.6*]. For the system $\Sigma = (F, G, H, J)$, the following are equivalent:
  (i) Σ is d-stabilizable;
  (ii) Σ is locally d-stabilizable;
  (iii) rank $[zI - \hat{F}(\omega) \quad \hat{G}(\omega)] = n$, $|z| \geq 1$, $\omega \in [0, 2\pi]$.

The result in Theorem 4.7 yields a procedure for computing a stabilizing feedback matrix since the matrix K_q given by (4.64) can be computed by recursively solving the RDE (4.63). The recursion must be carried out a sufficient number of steps until a stabilizing feedback is obtained. The computations could be performed by using the fast Fourier transform.

In contrast to the discrete-time case, there does not appear to be a recursive procedure based on the Riccati equation for solving for a stabilizing feedback in the continuous-time case. As a result, the above technique for showing that local d-stabilizability implies d-stabilizability cannot be carried over to the continuous-time case. However, we can use Delchamps' lemma [4.25], [4.22] to show that local stabilizability implies that the global versions of the ARE's (4.61) and (4.62) have solutions over $\ell^1(\mathbb{Z}, \mathbb{R})$, which in turn implies stabilizability. The details are as follows.

LEMMA (*Variation of Delchamps* [*4.22*]). *Suppose that Σ is locally c-stabilizable (resp., locally d-stabilizable). Then for each fixed $\omega \in [0, 2\pi]$, the entries of the solution P_ω to the ARE (4.61) (the ARE (4.62)) are analytic functions (given by Taylor series with real coefficients) of the entries of $\hat{F}(\omega)$, $\hat{F}(\omega)^*$, $\hat{G}(\omega)$, and $\hat{G}(\omega)^*$.*

Proof. The result follows by a minor modification of the work of Delchamps [4.22], who considers a matrix pair (F, G) with entries in the reals $\mathbb{R}$ only. When (F, G) have entries in $\mathbb{C}$, as is the case here, it is necessary to consider the conjugate transposes $\hat{F}(\omega)^*$ and $\hat{G}(\omega)^*$, i.e., the solution to the ARE is an analytic function of the entries of $\hat{F}(\omega)$, $\hat{G}(\omega)$ *and* the entries of $\hat{F}(\omega)^*$, $\hat{G}(\omega)^*$. $\square$

THEOREM 4.9. *Suppose that Σ is locally c-stabilizable (locally d-stabilizable). Then there exists a unique $n \times n$ matrix P over $\ell^1(\mathbb{Z}, \mathbb{R})$ such that*

$$\hat{P}(\omega) = P_\omega \quad \text{for all} \quad \omega \in [0, 2\pi], \tag{4.66}$$

where P_ω is the (p.d.) solution to the ARE (4.61) (the ARE (4.62)).

Proof. Suppose that Σ is locally c-stabilizable. Then by the Lemma and the Arens–Calderon theorem (see Bonsal and Duncan [4.26, pp. 105–

106]), it follows that there is an $n \times n$ matrix P over $\ell^1(\mathbb{Z}, \mathbb{R})$ which satisfies (4.66). Since the algebra $\ell^1(\mathbb{Z}, \mathbb{R})$ is semisimple, P is unique. A similar argument gives the discrete-time case. $\square$

Combining the above results, we have the following fundamental result.

THEOREM 4.10. The system Σ is c-stabilizable if and only if it is locally c-stabilizable, in which case,

$$K_c = G^*P \qquad (4.67)$$

is a stabilizing feedback, where $\hat{P}(\omega) = P_\omega =$ solution to the ARE (4.61). In addition, if the system Σ is locally d-stabilizable,

$$K_d = (G^*PG + I)^{-1}G^*PF \qquad (4.68)$$

is a stabilizing feedback, where $\hat{P}(\omega) = P_\omega =$ solution to the ARE (4.62).

The stabilizing feedbacks (4.67) and (4.68) are optimal. For instance, in the continuous-time case, the feedback control law

$$u(t, r) = -\sum_{j=-\infty}^{\infty} K_c(r - j)x(t, r),$$

where $K_c = G^*P$, minimizes (for each $\omega \in [0, 2\pi]$) the quadratic performance criterion

$$\int_0^\infty [\hat{x}(t, \omega)^* \hat{V}(\omega)\hat{x}(t, \omega) + \hat{u}(t, \omega)^* \hat{u}(t, \omega)] \, dt, \qquad (4.69)$$

where again * denotes the conjugate transpose. Here we are assuming that the input u_t and initial state x_0 have entries in $\ell^1(\mathbb{Z}, \mathbb{R})$ so that the Fourier transforms in (4.69) are well defined.

It is worth emphasizing that if all we want is a stabilizing feedback, in the discrete-time case it is not necessary to compute the solution P to the global ARE, since we can solve the RDE (4.63) recursively until we obtain a stabilizing feedback given by (4.64). Unfortunately, we do not have an analogous recursive procedure for computing a stabilizing feedback in the continuous-time case. However, we can first discretize in time using the techniques given in Section 4.3, and then we can apply the recursive procedure.

4.8. STABILIZATION BY DYNAMIC OUTPUT FEEDBACK

Let $\Sigma = (F, G, H, J)$ be a system over $\ell^1(\mathbb{Z}, \mathbb{R})$. To make the following development as simple as possible, throughout this section we assume that $J = 0$, and we write $\Sigma = (F, G, H)$. In this section we study the question as to when there exists a feedback system $\Sigma_{fb} = (A, B, C, D)$ over $\ell^1(\mathbb{Z}, \mathbb{R})$, where A is $\bar{n} \times \bar{n}$, B is $\bar{n} \times p$, C is $m \times \bar{n}$, D is $m \times p$, such that the resulting closed-loop system is internally stable. In the discrete-time case, the closed-loop system consisting of Σ and Σ_{fb} is given by the state equations

$$x_{k+1} = F_{*_r}x_k + G_{*_r}u_k$$
$$y_k = H_{*_r}x_k$$
$$\xi_{k+1} = A_{*_r}\xi_k + B_{*_r}y_k$$
$$u_k = -C_{*_r}\xi_k - D_{*_r}y_k + v_k.$$

Here ξ_k is the state of the feedback system Σ_{fb} and v_k is an external input. The above state equations for the closed-loop system can be written in the form

$$\begin{bmatrix} x_{k+1} \\ ---- \\ \xi_{k+1} \end{bmatrix} = \begin{bmatrix} F - GDH & \vdots & -GC \\ ---- & & ---- \\ BH & \vdots & A \end{bmatrix} \begin{bmatrix} x_k \\ -- \\ \xi_k \end{bmatrix} + \begin{bmatrix} G \\ -- \\ 0 \end{bmatrix} v_k \quad (4.70a)$$

$$y_k = [H \; \vdots \; 0] \begin{bmatrix} x_k \\ -- \\ \xi_k \end{bmatrix}. \quad (4.70b)$$

We shall say that Σ is *d-stabilizable by dynamic output feedback* if there exists a feedback system Σ_{fb} such that the closed-loop system given by (4.70a, b) is *d*-stable. There is an analogous notion of *c*-stabilizability by dynamic output feedback – We leave the details to the interested reader.

We will solve the output feedback stabilization problem by combining the results in Section 4.6 and 4.7 with a 'dualization' of the results in these two sections. We begin with the definition of the dual system and dual properties.

DEFINITION 4.6. The *dual system* Σ^* associated with the system $\Sigma = (F, G, H)$ is the system $\Sigma^* = (F^*, H^*, G^*)$, where as before * denotes the operation defined by (4.57). The system Σ is *coreachable* if the dual system Σ^* is reachable. The system Σ is *c-detectable* (*d-detectable*) if the dual system Σ^* is *c*-stabilizable (*d*-stabilizable). The system Σ is *locally c-detectable* (*locally d-detectable*) if the dual Σ^* is locally *c*-stabilizable (locally *d*-stabilizable).

All of the results in Sections 4.6 and 4.7 can be dualized in the obvious way to yield results on coreachability and detectability. For instance, dualizing Theorem 4.8, we have the following result.

PROPOSITION 4.8. *For the system* $\Sigma = (F, G, H)$, *the following are equivalent*:
  (i) Σ *is d-detectable;*
  (ii) Σ *is locally d-detectable;*
  (iii) *rank* $[zI \doteq \hat{F}(\omega)^* \; \hat{H}(\omega)^*] = n, \; |z| \geq 1, \; \omega \in [0, 2\pi];$
  (iv) *there exists a* $n \times p$ *matrix* L *over* $\ell^1(\mathbb{Z}, \mathbb{R})$ *such that* $F - LH$ *is d-stable.*

Proof. The equivalence of (i), (ii), and (iii) is immediate from Theorem 4.8. By definition, *d*-detectability is equivalent to the existence of a $p \times n$ matrix M over $\ell^1(\mathbb{Z}, \mathbb{R})$ such that $F^* - H^*M$ is *d*-stable. Since stability is not affected by the *-operation defined by (4.57), the matrix $(F^* - H^*M)^* = F - M^*H$ is *d*-stable. Thus, (iv) is equivalent to (i). $\square$

We then have the following result on the existence of a stabilizing dynamic compensator.

THEOREM 4.11. *Suppose that* Σ *is both locally d-stabilizable and locally d-detectable; that is*

$$rank[zI - \hat{F}(\omega) \quad \hat{G}(\omega)] = n, \; |z| \geq 1, \qquad \omega \in [0, 2\pi] \quad (4.71)$$

and

$$rank[zI - \hat{F}(\omega)^* \; \hat{H}(\omega)^*] = n, \; |z| \geq 1, \qquad \omega \in [0, 2\pi] \quad (4.72)$$

Then Σ *is d-stabilizable by dynamic output feedback.*

Proof. If (4.71) and (4.72) are satisfied, there exist matrices K and L over $\ell^1(\mathbb{Z}, \mathbb{R})$ such that $F - GK$ and $F - LH$ are *d*-stable. Now consider the feedback system $\Sigma_{fb} = (A, B, C, D)$ with $A = F - LH - GK$, $B = L$, $C = K$, and $D = 0$. From (4.70a), the system matrix F_{cl} of the resulting closed-loop system is given by

$$F_{cl} = \left[\begin{array}{c|c} F & -GK \\ \hline LH & F - LH - GK \end{array}\right].$$

Now let P denote the $2n \times 2n$ matrix defined by

$$P = \left[\begin{array}{c|c} I & 0 \\ \hline I & I \end{array}\right].$$

Then

$$P^{-1}F_{cl}P = \left[\begin{array}{c|c} F - GK & -GK \\ \hline 0 & F - LH \end{array}\right]$$

Clearly, the matrix $P^{-1}F_{cl}P$ is d-stable if and only if $F - GK$ and $F - LH$ are d-stable. In addition, since det $P = \pm 1$, F_{cl} is d-stable if and only if $P^{-1}F_{cl}P$ is d-stable. Hence the feedback system $\Sigma_{fb} = (F - LH - GK, L, K, 0)$ is stabilizing. $\square$

It turns out that the rank conditions (4.71) and (4.72) *are also necessary* for the existence of a dynamic feedback system which results in a d-stable closed-loop system. To see this, suppose that there is a feedback system $\Sigma_{fb} = (A, B, C, D)$ such that the closed-loop system given by (4.70a, b) is d-stable. Thus using the local criterion for d-stability, we have that

$$\text{rank} \left[\begin{array}{c|c} zI - \hat{F}(\omega) + \hat{G}(\omega)\hat{D}(\omega)\hat{H}(\omega) & \hat{G}(\omega)\hat{C}(\omega) \\ \hline -\hat{B}(\omega)\hat{H}(\omega) & zI - \hat{A}(\omega) \end{array}\right] =$$
$$n + \bar{n}, \; |z| \geq 1, \; \omega \in [0, 2\pi]. \quad (4.73)$$

Now if Σ is not locally d-stabilizable, there exists a nonzero $\gamma \in \mathbb{C}^n$ such that

$$\gamma'[z_0 I - \hat{F}(\omega_0) \quad \hat{G}(\omega_0)] = 0,$$

for some z_0 such that $|z_0| \geq 1$ and some $\omega_0 \in [0, 2\pi]$. Then

$$[\gamma' \mid 0] \left[\begin{array}{c|c} z_0 I - \hat{F}(\omega_0) + \hat{G}(\omega_0)\hat{D}(\omega_0)\hat{H}(\omega_0) & \hat{G}(\omega_0)\hat{C}(\omega_0) \\ \hline -\hat{B}(\omega_0)\hat{H}(\omega_0) & z_0 I - \hat{A}(\omega_0) \end{array}\right] = 0,$$

which contradicts (4.73). Thus, Σ is locally d-stabilizable. A similar argument shows that Σ is also locally d-detectable. $\square$

Conditions for the existence of a stabilizing dynamic feedback system can be expressed in terms of the system's transfer function matrix $W(z_1, z_2)$:

THEOREM 4.13. Let $W(z_1, z_2)$ be a strictly proper rational (in z_1) transfer function matrix. Then $W(z_1, z_2)$ has a realization $\Sigma = (F, G, H)$ which is d-stabilizable by dynamic output feedback if and only if $W(z_1, z_2)$ has a right polynomial matrix factorization

$$W(z_1, z_2) = P(z_1, z_2)Q(z_1, z_2)^{-1},$$

with

$$\text{rank}[Q(z, e^{i\omega})\quad P(z, e^{i\omega})] = m, |z| \geq 1, \quad \omega \in [0, 2\pi]. \quad (4.74)$$

The proof of Theorem 4.13 follows from existing results relating state representations and matrix-fraction descriptions of systems over a ring (e.g., see Khargonekar and Sontag [4.5]). We omit the details.

We should also mention that the stabilizability criterion (4.74) is similar to a stabilizability criterion derived by Guiver and Bose [4.27] for two-dimensional digital filters.

Given the transfer-function matrix $W(z_1, z_2)$, in order to test for stabilizability it is not necessary to compute a polynomial matrix factorization with the property (4.74). In particular, we have the following result which is an application of the results of Khargonekar and Sontag [4.5, Theorem 6.15].

THEOREM 4.14. Let $W(z_1, z_2)$ be a strictly proper rational (in z_1) transfer function matrix whose coefficients belong to either the PID $\mathbb{R}[z_2, z_2^{-1}]$ or the PID $\mathbb{R}_{\bar{7}}(z_2)$. (See Section 4.6 for the definition of these PID's.) Then $W(z_1, z_2)$ has a realization Σ which is d-stabilizable by dynamic output feedback if and only if the canonical realization of $W(z_1, z_2)$ is locally d-detectable.

The stabilizability test in Theorem 4.14 is constructive since there are procedures for computing canonical (reachable and observable) realizations of a transfer-function matrix with coefficients in a PID. For example, see Rouchaleau and Sontag [4.28].

4.9. APPLICATION TO TRACKING

As an application of the stabilization criteria derived in the previous sections, in this section we consider the problem of tracking with internal stability.

Let $\Sigma = (F, G, h)$ be a m-input single-output system over $\ell^1(\mathbb{Z}, \mathbb{R})$, which is interpreted as a spatially-distributed discrete-time system given by the state equations

$$x_{k+1} = F *_r x_k + G *_r u_k \qquad (4.75a)$$

$$y_k = h *_r x_k. \qquad (4.75b)$$

We want to construct a feedback system so that the resulting closed-loop system is d-stable and so that the output y_k tracks a given reference signal γ_k (i.e., we want $\sup_{r \in \mathbb{Z}} |y(k, r) - \gamma(k, r)| \to 0$ as $k \to \infty$). We assume that the reference signal γ_k is given by the following qth-order difference equation

$$\gamma_{k+q} = \sum_{j=0}^{q-1} a_j *_r \gamma_{k+j}, \qquad k \geq 0,$$

where the a_j are fixed elements of $\ell^1(\mathbb{Z}, \mathbb{R})$ and the initial conditions γ_0, $\gamma_1, \ldots, \gamma_{q-1}$ are arbitrary elements of $\ell^\infty(\mathbb{Z}, \mathbb{R})$.

Letting $e_k = y_k - \gamma_k$ denote the error, we construct the *error-driven system* given by the equation

$$\eta_{k+1} = A *_r \eta_k + b *_r e_k, \qquad (4.76)$$

where

$$A = \begin{bmatrix} 0 & 0 & 0 & \ldots & a_0 \\ 1 & 0 & 0 & \ldots & a_1 \\ 0 & 1 & 0 & \ldots & a_2 \\ \vdots & & & \ddots & \vdots \\ 0 & 0 & 0 & \ddots & 1 & a_{q-1} \end{bmatrix}, \qquad b = \begin{bmatrix} 1 \\ 0 \\ \vdots \\ 0 \end{bmatrix}.$$

Now combine (4.75a, b) and (4.76), which gives the *augmented system*

$$\begin{bmatrix} x_{k+1} \\ \hdashline \eta_{k+1} \end{bmatrix} = \begin{bmatrix} F & \vdots & 0 \\ \hdashline bh & \vdots & A \end{bmatrix} \begin{bmatrix} x_k \\ \hdashline \eta_k \end{bmatrix} + \begin{bmatrix} G \\ \hdashline 0 \end{bmatrix} u_k - \begin{bmatrix} 0 \\ \hdashline b \end{bmatrix} \gamma_k$$

$$y_k = h x_k.$$

Define

$$
F_a = \left[\begin{array}{c|c} F & 0 \\ \hline bh & A \end{array}\right], \qquad G_a = \left[\begin{array}{c} G \\ \hline 0 \end{array}\right].
$$

Now suppose that the given system $\Sigma = (F, G, h)$ is locally d-detectable and that the augmented system given by (F_a, G_a, h) is locally d-stabilizable. Then by the results in Section 4.7, there exists a n-element column vector L over $\ell^1(\mathbb{Z}, \mathbb{R})$ such that $F - Lh$ is d-stable, and there exists a $m \times (n + q)$ matrix K over $\ell^1(\mathbb{Z}, \mathbb{R})$ such that $F_a - G_a K$ is d-stable. Partition K so that $K = [K_1\ K_2]$, where K_1 is a $m \times n$ matrix and K_2 is a $m \times q$ matrix, and consider the following feedback system

$$
\xi_{k+1} = (F - GK_1 - Lh)\xi_k - GK_2\eta_k + Ly_k \tag{4.77a}
$$

$$
u_k = -K_1\xi_k - K_2\eta_k. \tag{4.77b}
$$

The resulting closed-loop system consisting of (4.75a, b), (4.76), and (4.77a, b) is given by the state equation

$$
\left[\begin{array}{c} x_{k+1} \\ \hline \eta_{k+1} \\ \hline \xi_{k+1} \end{array}\right] = \left[\begin{array}{c|c|c} F & -GK_2 & -GK_1 \\ \hline bh & A & 0 \\ \hline Lh & -GK_2 & F - GK_1 - Lh \end{array}\right] \left[\begin{array}{c} x_k \\ \hline \eta_k \\ \hline \xi_k \end{array}\right] - \left[\begin{array}{c} 0 \\ \hline b \\ \hline 0 \end{array}\right] y_k. \tag{4.78}
$$

Now let P denote the $(2n + q) \times (2n + q)$ matrix given by

$$
P = \left[\begin{array}{c|c|c} I_n & 0 & 0 \\ \hline 0 & I_q & 0 \\ \hline I_n & 0 & I_n \end{array}\right],
$$

where $I_n(I_q)$ is the $n \times n(q \times q)$ identity matrix. Then letting F_b denote the $(2n + q) \times (2n + q)$ system matrix in the state equation (4.78), we have that

$$
P^{-1}F_b P = \left[\begin{array}{c|c|c} F - GK_1 & -GK_2 & -GK_1 \\ \hline bh & A & 0 \\ \hline 0 & 0 & F - Lh \end{array}\right]. \tag{4.79}
$$

Since $F_a - G_a[K_1 \ K_2]$ is d-stable and $F - Lh$ is d-stable, from the block-triangle form of (4.79) we have that $P^{-1}F_bP$ is also d-stable. Thus, the closed-loop system given by (4.78) is d-stable. Finally, it is not difficult to show that the transfer function from the reference input γ_k to the error e_k is of the form $a(z, \omega)/b(z, \omega)$, where $b(z, \omega) \neq 0$, $|z| \geq 1$, $\omega \in [0, 2\pi]$, and where $a(z, \omega)$ is divisible by $z^q - \Sigma_j \hat{a}_j(\omega)z^j$. Then using the definition of the reference γ_k, we have that

$$\sup_{r\in Z} |y(k, r) - \gamma(k, r)| \to 0 \quad \text{as} \quad k \to \infty$$

for *any* initial states $x(0, \ .) \in \ell^\infty(Z, \mathbb{R}^n)$, $\xi(0, \ .) \in \ell^\infty(Z, \mathbb{R}^q)$. Thus, the error-driven system (4.76) and the feedback system (4.77a, b) solve the tracking problem posed above.

EXAMPLE 4.6. Consider the seismic cable with the following parameter values (taken from El-Sayed and Krishnaprasad [4.3]):

$$k = 8, \qquad c = 10, \qquad M = 40.$$

With these parameter values and with the sampling interval $T = 1$, the coefficient matrices $F_{1,d}(0)$, $F_{1,d}(1)$, and $G_{1,d}(0)$ comprising the discretization (4.39) are given by (see (4.40), (4.41), (4.42))

$$F_{1,d}(0) = \begin{bmatrix} .90933 & .85563 \\ -.17112 & .69543 \end{bmatrix}$$

$$F_{1,d}(1) = \begin{bmatrix} .04533 & 0 \\ .08556 & 0 \end{bmatrix}$$

$$G_{1,d}(0) = \begin{bmatrix} .01133 \\ .02139 \end{bmatrix}$$

Letting $\hat{F}_d(\omega)$ and $\hat{G}_d(\omega)$ denote the Fourier transforms of the coefficient matrices of this discretization, we have

$$\hat{F}_d(\omega) = \begin{bmatrix} .90933 + .09066 \cos \omega & .85563 \\ -.17112(1 - \cos \omega) & .69543 \end{bmatrix}$$

$$\hat{G}_d(\omega) = \hat{G}_d = \begin{bmatrix} .01133 \\ .02139 \end{bmatrix}.$$

Now

$$\det [\hat{G}_d \quad \hat{F}_d(\omega)\hat{G}_d] = -.00047, \quad \text{all} \quad \omega \in [0, 2\pi].$$

Hence, the discretization is locally reachable, and by Theorem 4.4, it is also reachable. With $H = [1 \quad 0]$ = coefficient of the output equation of the cable system (see Example 4.2), we have

$$\det [\hat{H}^* \quad \hat{F}_d(\omega)^*\hat{H}^*] = .85563, \quad \text{all} \quad \omega \in [0, 2\pi].$$

Thus, by the dual of Theorem 4.4, the discretization is coreachable. We can therefore place the poles of this discretization by employing dynamic output feedback as defined in Section 4.8. Now suppose that we want the discretized output $y(k, r)$ to track a step function $\gamma(k, r) = \theta(r), r \in \mathbb{Z}$, $k \geq 0$, where $\theta \in \ell^\infty(\mathbb{Z}, \mathbb{R})$. Then the model for the step reference is

$$\gamma_{k+1} = \gamma_k \quad \text{with initial condition} \quad \gamma_0 = \theta,$$

and hence the state equation (4.76) for the error-driven system becomes

$$\eta_{k+1} = \eta_k + e_k.$$

Letting $\hat{F}_a(\omega)$ and $\hat{G}_a(\omega)$ denote the Fourier transforms of the coefficient matrices of the augmented system consisting of the discretized cable system and the error-driven system, we have

$$\hat{F}_a(\omega) = \begin{bmatrix} .90933 + .09066 \cos \omega & .85563 & 0 \\ -.17112(1 - \cos \omega) & .69543 & 0 \\ 1 & 0 & 1 \end{bmatrix}$$

$$\hat{G}_a(\omega) = \hat{G}_a = \begin{bmatrix} .01133 \\ .02139 \\ 0 \end{bmatrix}$$

Now since the discretized cable system is coreachable, we have that

$$\text{rank} [zI - \hat{F}_a(\omega) \quad \hat{G}_a] = 3 \quad \text{for} \quad z \neq 1 \quad \text{and} \quad \omega \in [0, 2\pi].$$

An easy check shows that

$$\text{rank} [I - \hat{F}_a(\omega) \quad \hat{G}_a] = 3 \quad \text{for all} \quad \omega \in [0, 2\pi].$$

It follows that the augmented system is reachable, and thus by using a

feedback system of the form (4.77a, b), we can track the step function *and*
we can place the poles of the closed-loop system. The coefficient matrices
of the feedback system could be computed using standard methods for
placing the poles of a single-input reachable system, or they could be
computed by solving the Riccati difference equation defined in Section
4.7. We leave the details to the interested reader.

ACKNOWLEDGEMENT

It is a pleasure to acknowledge the U.S. Army Research Office for their
support of this work which was carried out under Contract No.
DAAG29–81–K–0166. The writer also wishes to thank P. P.
Khargonekar for several helpful comments in the preparation of this
chapter.

REFERENCES

[4.1] S. M. Melzer and B. C. Kuo, 'Optimal Regulation of systems described by a
       countably infinite number of objects', *Automatica*, Vol. 7, pp. 359–366, 1971.

[4.2] K. C. Chu, 'Optimal Decentralized Regulation for a String of Coupled Systems',
       IEEE Trans. Automatic Control, Vol. AC-19, pp. 243–246, 1974.

[4.3] M. L. El-Sayed and P. S. Krishnaprasad, 'Homogeneous Interconnected Systems:
       An Example', *IEEE Trans. Automatic Control*, Vol. AC-26, pp. 894–901, 1981.

[4.4] W. L. Green and E. W. Kamen, 'Stabilizability of Linear Systems over a Com-
       mutative Normed Algebra with Applications to Spatially-distributed and Parameter-
       dependent Systems', original version submitted for publication in July 1981, to
       appear in Siam, J. Control & Optimization, January 1985.

[4.5] P. P. Khargonekar and E. D. Sontag, 'On the Relation between Stable Matrix
       Fraction Factorizations and Regulable Realizations of Linear Systems over Rings',
       IEEE Trans. Automatic Control, Vol. AC-27, pp. 627–638, 1982.

[4.6] E. W. Kamen and P. P. Khargonekar, 'On the Control of Linear Systems whose
       Coefficients Are Functions of Parameters', *IEEE Trans. Automatic Control*, Vol.
       AC-29, pp. 25–33, January 1984.

[4.7] N. Dunford and J. T. Schwartz, 'Linear Operators Part I: General Theory', Vol. VII
       in *Pure and Applied Math. Series*, Interscience, 1971.

[4.8] E. D. Sontag, 'Linear Systems over Commutative Rings: A Survey', *Ricerche di
       Automatica*, Vol. 7, pp. 1–34, 1976.

[4.9] R. E. Edwards, 'Fourier Series, A Modern Introduction', Volume 1, *Graduate Texts
       in Math.*, Vol. 64, Springer-Verlag, New York, 1979 (second edition).

[4.10] M. Hazewinkel, 'Moduli and Canonical Forms for Linear Dynamical Systems, II:
       The Topological Case', *Math. System Theory*, Vol. 10, pp. 363–385, 1977.

[4.11] M. Hazewinkel, 'On Families of Linear systems: Degeneration Phenomena', in *Algebraic and Geometric Methods in Linear Systems Theory, Lectures in Applied Mathematics*, Vol. 18, C. I. Byrnes and C. Martin (eds.), American Math. Society, Providence, pp. 157–189, 1980.

[4.12] M. Hazewinkel, '(fine) moduli (spaces): What are they, and what are they good for?', in *Geometrical Methods for Linear Systems Theory*, C. I. Byrnes and C. Martin (eds.), D. Reidel Publ. Co., Dordrecht, The Netherlands, 1980.

[4.13] C. I. Byrnes, 'On the Control of Certain Deterministic, Infinite-dimensional Systems by Algebro-geometric Techniques', *American Journal of Math.*, Vol. 100, pp. 1333–1381, 1978.

[4.14] A. Tannenbaum, 'Invariance and System Theory: Algebraic and Geometric Aspects', *Lectures Notes in Mathematics*, Vol. 845, Springer-Verlag, Berlin 1981.

[4.15] E. W. Kamen and W. L. Green, 'Asymptotic Stability of Linear Difference Equations Defined over a Commutative Banach Algebra', *Journal Math. Analysis and Applications*, Vol. 75, pp. 584–601, 1980.

[4.16] D. Goodman, 'Some Stability Properties of Two-dimensional Linear Shift-invariant Digital Filters', *IEEE Trans. Circuits and Systems*, Vol. CAS-24, pp. 201–208, 1977.

[4.17] S. K. Berberian 'Lectures in Functional Analysis and Operator Theory', *Graduate Texts in Math.*, Vol. 15, Springer-Verlag, New York, 1974.

[4.18] M. L. J. Hautus, 'Stabilization Controllability and Observability of Linear Autonomous Systems', *Ned. Akad. Wetenschappen, Proc.*, Ser. A, Vol. 73, pp. 448–455, 1970.

[4.19] D. L. Kleinman, 'An Easy Way to Stabilize a Linear Constant System', *IEEE Trans. Automatic Control*, Vol. AC-15, p. 692, 1970.

[4.20] C. I. Byrnes, 'On the stabilizability of Linear Control Systems Depending on Parameters', in *Proc. 18th IEEE Conference Decision and Control*, Ft. Lauderdale, pp. 233–236, 1979.

[4.21] C. I. Byrnes, 'Algebraic and Geometric Aspects of the Analysis of Feedback Systems', in *Geometrical Methods for the Theory of Linear Systems*, C. I. Byrnes and C. Martin (eds.), D. Reidel Publ. Co., Dordrecht, The Netherlands, pp. 85–124, 1980.

[4.22] D. F. Delchamps, 'Analytic Stabilization and the Algebraic Riccati Equation', in *Proc. 22nd IEEE Conference Decision and Control*, San Antonio, pp. 1396–1401, 1983.

[4.23] A. S. Morse, 'Ring Models for Delay Differential Systems', *Automatica*, Vol. 12, pp. 529–531, 1976.

[4.24] H. Kwakernaak and R. Sivan, *Linear Optimal Control Systems*, Wiley, New York, 1972.

[4.25] D. F. Delchamps, 'A Note on the Analyticity of the Riccati Metric', in *Algebraic and Geometric Methods in Linear Systems Theory, Lectures in Applied Mathematics*, Vol. 18, C. I. Byrnes and C. Martin, (eds.), American Math. Society, Providence, pp. 37–41, 1980.

[4.26] F. F. Bonsal and J. Duncan, *Complete Normed Algebras, Ergebnisse der Mathematik und ihrer Grenzgebiete*, Vol. 80, Springer-Verlag, New York, 1973.

[4.27] J. P. Guiver and N. K. Bose, 'Causal and Weakly causal 2–D Filters with Applications in Stabilization', in *Multi-dimensional Systems Theory*, N. K. Bose (ed.), D. Reidel Publ. Co., Dordrecht, The Netherlands, pp. 52–100 (this volume), 1984.

[4.28] Y. Rouchaleau and E. D. Sontag, 'On the existence of Minimal Realizations of Linear Dynamical Systems over Noetherian Integral Domains', *Journal Computer and Syst. Sci.*, Vol. 18, pp. 65–75, 1978.

Chapter 5

H. M. Valenzuela and N. K. Bose

Linear Shift-Variant Multidimensional Systems

5.1. INTRODUCTION

The study of 1–D time-varying systems, now well-documented in the literature, was motivated to a great extent by the need to design adaptive control systems, where the parameters of the controller are adjusted to counterbalance the fluctuations in process dynamics stemming from changes in environment. For example, variations of flight conditions (including flight speed) of supersonic aircrafts and missiles during rapid ascent through the atmosphere introduce time-varying parameter variations. In space flight, relevant problems that had to be tackled included those [5.1] of transferring a space vehicle from one orbit to another, rendezvous and interplanetary guidance. The development of the rigid-body equations of motion of artificial satellites requires accounting for cyclic variations of inertial torques that occur as the vehicle progresses in its orbit. Similarly, successful solution of the satellite rendezvous problem requires dealing with the combined complexity of both mass and orbital variations; for with both the target and interceptor in orbit, the interceptor must expend fuel-mass in maneuvering; further, the kinematics of interceptor guidance introduce additional time-varying parameters forcing the characterizing differential equations to have both periodic and aperiodic coefficients. Also of interest is a class of problems which relate to the gyroscopic stabilization of orbiting satellites. In circuit and systems theory specific applications concerned with the theory of modulators, parametric amplifiers and harmonic generators use linear circuits with periodically varying parameters [5.2].

   The developed tools of multidimensional systems theory, as documented in [5.3], provide scopes for considerable applications in a broad range of problems. However, over the last few years, it is fair to say that the theory of linear shift-invariant (LSI) two or more dimensional systems have been well understood. However, with the increased activities in the area of image processing (optical and digital) many physical problems are characterizable by a 2–D linear integral operator,

147

N. K. Bose (ed.), Multidimensional Systems Theory, 147–183
© 1985 by D. Reidel Publishing Company.

$$g(x, y) = \int_{-x}^{x} \int_{-x}^{x} f(u, v)h(x, y; u, v) \, dv$$

where $g(x, y)$, $f(x, y)$ are, respectively, the output and input while
$h(x, y; u, v)$ is the system response at (x, y) to an unit impulse applied at
(u, v). The presence of additive noise in the preceding input-output
model of a physical process could also be expected. In 1977, Goodman
stated that "the type of coherent optical processing which is by far the least
explored to date is that of linear space-variant filtering" [5.4], and the
importance of overcoming this deficiency was realized because space-
variant processing is required in various optical and digital data processing applications including restoration of images degraded either by
space-variant aberrations or space-variant motion blur, restoration of
radio-graphs blurred by a space-variant source penumbra, restoration of
rotation blur of 2–D patterns where different objects suffer from different
blur according to their distance from the center of rotation and also their
angular position [5.5], and performance of particular transformations
(like the Mellin transform and Abel transform). Also, emerging applications in areas like robotic vision, have necessitated the development of
techniques suitable for recognition, patterns, which may be subjected to
the effects of unknown rotation and scaling. One such technique is based
on the use of scaled transform (a modified Fourier transform), which
has been shown to be equivalent to a class of linear shift-variant (LSV)
filters, referred to as the class of 'form-invariant' filters [5.6].

Actually, the problem of 2–D images degraded by spatially varying
point spread functions has been tackled since about the early seventies.
Sawchuk [5.7] transformed the spatially varying problem to a spatially
invariant one via coordinate transformations and as a result some
generality in the approach had to be sacrificed, especially with the
presence of additive noise. In that situation, however, statistics of the
noise and object random processes could change under a geometrical
coordinate transformation. The usual assumption of the minimum mean-
square error estimation (MMSE) method is that these statistics are jointly
stationary, and this may not be valid in many cases. Frieden [5.8]
developed a positive (an incoherent object scene is a spatial radiance
distribution and cannot be negative) restoring formula from a statistical
communication theory model of image formation by which the optimal
restored object from a set of candidate objects is required to obey a
principle of maximum entropy. Frieden's approach, though sufficiently
general, poses some dimensionality problems in implementation, in spite

of the fact that the restorations besides being positive are spatially smooth and not overly sensitive to noise in the image data at object points that are near the background radiance. Then, there is the class of iterative methods based on gradient type [5.9] and conjugate gradient type [5.10] of optimization procedures. Iterative methods are not very suited for image restoration because of the problem of noise suppression. If the blurred image to be processed includes a noise, then this noise strongly deteriorates the quality of the restored image as the number of iterations increase. Thus, a method is desired that enables one to suppress noise amplification more efficiently, while restoring the image sharply. The three ways to suppress noise amplification are: (1) stop the iterations at a moderate iteration number; (2) introduce constraints; and (3) reblur the blurred image as done in [5.11] for LSI systems. In spite of these efforts, the general problem of space-time computational complexity has limited the use of practically all methods (including those already cited and others [5.12], [5.13]), in shift-variant multidimensional problems.

A more general approach than that permitted via use of coordinate transformation was pursued recently in [5.14], where the analysis was restricted to one-dimensional (1–D) discrete LSV systems. Investigations, however, were carried out both in the frequency (using the discrete version of Zadeh's generalized transfer function introduced in the discussion of continuous time systems [5.15]) and time (where the impulse response sequence was assumed to be either expressible or approximated in k-th order degenerate form) domains. In this chapter attention will be focused on the development of a state-space model for a k–D LSV system whose impulse response is expressible in K-th order degenerate form. For convenience in exposition, the development of the model will be initiated for the 2–D quarter-plane causal case and a complete proof for the feasibility of extension to the k–D ($k > 2$) case based on a $(2^k - 1)$-point recursion will be given. The possibility of generalizing the model to cover multi-dimensional weakly causal LSV systems will be substantiated. The model for the inverse system will be derived and applied to problems in deconvolution. Nontrivial physically motivated examples will be included.

A two dimensional discrete LSV system can be described by the discrete sum

$$y(n_1, n_2) = \sum_{m_1 = -\infty}^{\infty} \sum_{m_2 = -\infty}^{\infty} h(n_1, n_2; m_1, m_2)u(m_1, m_2), \quad (5.1)$$

where $u(\mathbf{m})$ is the input at coordinate point $\mathbf{m} = (m_1, m_2)$, $y(\mathbf{n})$ is the output at coordinate point $\mathbf{n} = (n_1, n_2)$ and $h(n_1, n_2; m_1, m_2)$ is the response of the system at the point $\mathbf{n}$ to a unit impulse at the point $\mathbf{m}$.

The input-output relation in (5.1) is far from being very convenient because it is not recursive and, therefore, the amount of computational time required for implementing the relation will be very large. In the case of 2–D discrete Linear Shift-Invariant systems the sum in (5.1) gets transformed into the standard 2–D convolution and there are very efficient techniques for calculating the 2–D discrete convolution. These techniques include transform domain methods that make use of fast algorithms and there are also state-space models for recursive implementation in the spatial domain [Chapter 4 in 5.3].

For LSV systems; the transform domain techniques cannot be used in general and, in addition to this, there are, as yet no state-space recursive models for 2–D LSV systems. Some very restrictive 1–D state-space models for LSV systems have appeared in the literature. They permit the analysis of 1–D LSV systems under very strong restrictions imposed on the impulse response [5.16], [5.17].

5.2. 2–D QUARTER PLANE CAUSAL STATE-SPACE MODEL

In order to develop a state-space model, the following definition has to be introduced as a natural extension of the 1–D case [5.14].

DEFINITION 5.1. A sequence $\{(h(n_1, n_2; m_1, m_2)\}$ is a K-th order degenerate sequence if it can be expressed as

$$h(n_1, n_2; m_1, m_2) = \sum_{i=1}^{K} \alpha_i(n_1, n_2)\beta_i(m_1, m_2), \qquad (5.2)$$

where $\alpha_i(n_1, n_2)$ $(i = 1, 2, \ldots, K)$ and $\beta_i(m_1, m_2)$ $(i = 1, 2, \ldots, K)$ are, respectively, linearly independent functions of (n_1, n_2) and (m_1, m_2).

Let $h(n_1, n_2; m_1, m_2)$ be the impulse response of a 2–D first quadrant quarter-plane causal discrete LSV system, and let $u(n_1, n_2)$ be the input to the system with support in the first quadrant. Then,

$$h(n_1, n_2; m_1, m_2) = 0 \quad \text{for} \quad n_1 < m_1 \quad \text{or} \quad n_2 < m_2, \quad (5.3)$$

and

$$u(n_1, n_2) = 0 \quad \text{for} \quad n_1 < 0 \quad \text{or} \quad n_2 < 0. \quad (5.4)$$

Substituting (5.3) and (5.4) in (5.1) the input-output relation is given by

$$y(n_1, n_2) = \sum_{m_1 = 0}^{n_1} \sum_{m_2 = 0}^{n_2} h(n_1, n_2; m_1, m_2)u(m_1, m_2). \quad (5.5)$$

If the impulse response $h(n_1, n_2; m_1, m_2)$ is a K-th order degenerate sequence given by (5.2), then, substituting (5.2) in (5.5) we have

$$y(n_1, n_2) = \sum_{m_1 = 0}^{n_1} \sum_{m_2 = 0}^{n_2} \left(\sum_{i = 1}^{K} \alpha_i(n_1, n_2)\beta_i(m_1, m_2) \right) u(m_1, m_2). \quad (5.6)$$

After some manipulations involving the interchanging of summations, we obtain

$$y(n_1, n_2) = \sum_{i = 1}^{K} \alpha_i(n_1, n_2)x_i(n_1, n_2), \quad (5.7)$$

where

$$x_i(n_1, n_2) = \sum_{m_1 = 0}^{n_1} \sum_{m_2 = 0}^{n_2} \beta_i(m_1, m_2)u(m_1, m_2). \quad (5.8)$$
$$i = 1, 2, \ldots, K$$

Let us consider $x_i(n_1, n_2)$ for an arbitrary but fixed i, $i = 1, 2, \ldots, K$. Splitting the summation in (5.8) according to the masks in Figure 5.1, we have

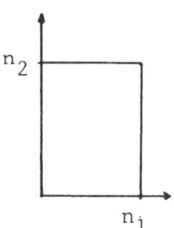

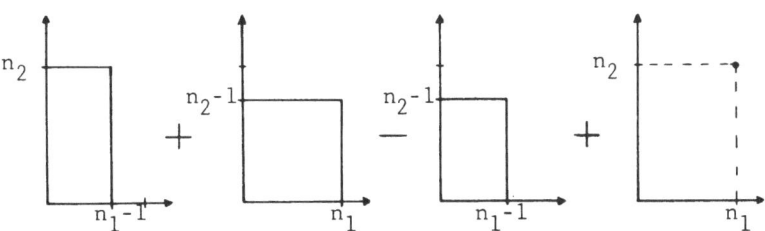

Fig. 5.1   Decomposition of rectangular grid for obtaining recursion.

$$x_i(n_1, n_2) = \sum_{m_1=0}^{n_1-1} \sum_{m_2=0}^{n_2} \beta_i(m_1, m_2)u(m_1, m_2) +$$

$$+ \sum_{m_1=0}^{n_1} \sum_{m_2=0}^{n_2-1} \beta_i(m_1, m_2)u(m_1, m_2) - \qquad (5.9)$$

$$- \sum_{m_1=0}^{n_1-1} \sum_{m_2=0}^{n_2-1} \beta_i(m_1, m_2)u(m_1, m_2) +$$

$$+ \beta_i(n_1, n_2)u(n_1, n_2).$$

Substituting (5.8) in (5.9) and making use of (5.7), we have

$$x_i(n_1, n_2) = x_i(n_1 - 1, n_2) + x_i(n_1, n_2 - 1) -$$

$$- x_i(n_1 - 1, n_2 - 1) + \beta_i(n_1, n_2)u(n_1, n_2)$$

$$i = 1, 2, \ldots, K. \qquad (5.10)$$

$$y(n_1, n_2) = \sum_{i=1}^{K} \alpha_i(n_1, n_2)x_i(n_1, n_2). \qquad (5.11)$$

Equations (5.10) and (5.11) can be rewritten as

$$\mathbf{x}(n_1, n_2) = \mathbf{x}(n_1 - 1, n_2) + \mathbf{x}(n_1, n_2 - 1) -$$

$$- \mathbf{x}(n_1 - 1, n_2 - 1) + \mathbf{b}(n_1, n_2)u(n_1, n_2), \quad (5.12)$$

$$\mathbf{y}(n_1, n_2) = \mathbf{c}(n_1, n_2)\mathbf{x}(n_1, n_2), \qquad (5.13)$$

where

$$\mathbf{x}(n_1, n_2) = \begin{bmatrix} x_1(n_1, n_2) \\ x_2(n_1, n_2) \\ \vdots \\ x_K(n_1, n_2) \end{bmatrix}, \qquad (5.14)$$

$$\mathbf{b}(n_1, n_2) = \begin{bmatrix} \beta_1(n_1, n_2) \\ \beta_2(n_1, n_2) \\ \vdots \\ \beta_K(n_1, n_2) \end{bmatrix}, \qquad (5.15)$$

$$\mathbf{c}(n_1, n_2) = [\alpha_1(n_1, n_2), \alpha_2(n_1, n_2), \ldots, \alpha_K(n_1, n_2)]. \qquad (5.16)$$

The initial conditions for the model above are:

$$\mathbf{x}(n_1, -1) = \mathbf{0}, \qquad \mathbf{x}(-1, n_2) = \mathbf{0} \qquad n_1, n_2 = 0, 1, 2, \ldots.$$

$$(5.17)$$

If the input array $\{u(n_1, n_2)\}$ has a finite support

$$S = \{(n_1, n_2), \quad 0 \le n_1 \le N_1, \quad 0 \le n_2 \le N_2\}, \quad (5.18)$$

and we are interested in computing the output $y(n_1, n_2)$ on S, then we only require a finite set of boundary conditions, which is given by

$$\mathbf{x}(n_1, -1) = \mathbf{0} \quad \text{for} \quad n_1 = 0, 1, \ldots, N_1, \quad (5.19)$$

$$\mathbf{x}(-1, n_2) = \mathbf{0} \quad \text{for} \quad n_2 = 0, 1, \ldots, N_2. \quad (5.20)$$

The state-space model introduced in (5.12)–(5.16) is based on a three term recursion, which means that in order to generate the state vector $\mathbf{x}(n_1, n_2)$ at a point (n_1, n_2) we need to have previously computed the state vectors at the points $(n_1 - 1, n_2)$, $(n_1, n_2 - 1)$ and $(n_1 - 1, n_2 - 1)$ in addition to the current input. See Figure 5.2.

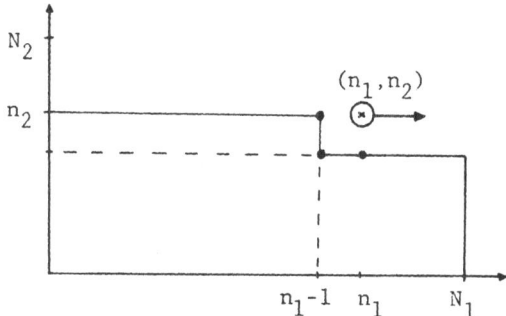

Fig. 5.2   Neighbors of (n_1, n_2) required to compute state at (n_1, n_2) by the state-space model.

From (5.10), it is easy to see that the states are decoupled; this is a nice feature which permits parallel computation of the states.

The state-space model introduced here has a structure similar to the Fornasini-Marchesini state-space representation [5.3]. The main difference between these two models, is that in (5.12) and (5.13) some of the matrices are non-constant.

The state-space model in (5.12) to (5.16) is not a unique state-space model representation of the system.

Let $\hat{\mathbf{x}}(n_1, n_2)$ be a new state vector defined by

$$\hat{\mathbf{x}}(n_1, n_2) = \mathbf{T}(n_1, n_2)\mathbf{x}(n_1, n_2). \quad (5.21)$$

Let the transformation matrix $\mathbf{T}\ (n_{1-}\ n_2)$ be nonsingular for all points

(n_1, n_2) in the region S, defined in (5.18). Making use of this transformation, Equations (5.12) to (5.16), become

$$\hat{x}(n_1, n_2) = \hat{A}_1(n_1, n_2)\hat{x}(n_1 - 1, n_2) +$$
$$+ \hat{A}_2(n_1, n_2)\hat{x}(n_1, n_2 - 1) - \qquad (5.22)$$
$$- \hat{A}_3(n_1, n_2)\hat{x}(n_1 - 1, n_2 - 1) + \hat{b}(n_1, n_2)u(n_1, n_2),$$

$$y(n_1, n_2) = \hat{c}(n_1, n_2)\hat{x}(n_1, n_2), \qquad (5.23)$$

where

$$\hat{A}_1(n_1, n_2) = T(n_1, n_2) \cdot T^{-1}(n_1 - 1, n_2), \qquad (5.24)$$

$$\hat{A}_2(n_1, n_2) = T(n_1, n_2) \cdot T^{-1}(n_1, n_2 - 1), \qquad (5.25)$$

$$\hat{A}_3(n_1, n_2) = T(n_1, n_2) \cdot T^{-1}(n_1 - 1, n_2 - 1), \qquad (5.26)$$

$$\hat{b}(n_1, n_2) = T(n_1, n_2)b(n_1, n_2), \qquad (5.27)$$

$$\hat{c}(n_1, n_2) = c(n_1, n_2)T^{-1}(n_1, n_2). \qquad (5.28)$$

From the above, we can say that the state-space representation in (5.22) to (5.28) is related to the state-space model in (5.12) to (5.16) by the transformation matrix in (5.21).

5.3. k–D STATE-SPACE MODEL

Now, the results obtained in Section 5.2 will be extended to the k-dimensional ($k > 2$) case. The state-space model in Section 5.2 was obtained by making use of a geometric argument which consists of splitting a 2–D grid into its components and summing over them in Equation (5.9). This derivation, even though simple, cannot be easily extended to higher dimensions. In this section, the particular structure of a matrix will be used to our advantage in obtaining a k–D recursive model. First, this will be done for the case when the impulse response sequence has support in a positive cone and, subsequently, weakly causal LSV systems will be tackled.

5.3.1. k–D Positive Cone Causal State-Space Model

A k–D positive cone causal discrete LSV system can be characterized by the superposition sum

$$y(n_1, \ldots, n_k) = \sum_{m_1 = 0}^{n_1} \cdots \sum_{m_k = 0}^{n_k}$$

$$h(n_1, \ldots, n_k; m_1, \ldots, m_k)u(m_1, \ldots, m_k),$$

(5.29)

where $u(\mathbf{m})$ is the input at coordinate point $\mathbf{m} = (m_1, m_2, \ldots, m_k)$, $y(\mathbf{n})$ is the output at coordinate point $\mathbf{n} = (n_1, n_2, \ldots, n_k)$ and $h(\mathbf{n}, \mathbf{m})$ is the response of the system at the point $\mathbf{n}$ to a unit impulse at the point $\mathbf{m}$. At this point, a definition, which is a natural extension of Definition 5.1, is given.

DEFINITION 5.2. A sequence $\{h(n_1, n_2, \ldots, n_k; m_1, m_2, \ldots, m_k)\}$, is K-th order degenerate if it can be expressed as

$$h(n_1, \ldots, n_k; m_1, \ldots, m_k) = \sum_{i = 1}^{K}$$

$$\alpha_i(n_1, \ldots, n_k)\beta_i(m_1, \ldots, m_k),$$

(5.30)

where $\alpha_i(n_1, \ldots, n_k)$ $(i = 1, 2, \ldots, K)$ and $\beta_i(m_1, \ldots, m_k)$ $(i = 1, 2, \ldots, K)$ are, respectively, linearly independent functions of $(n_1, n_2, \ldots, n_k)$ and $(m_1, m_2, \ldots, m_k)$.

Substituting (5.30) in (5.29), after carrying out some manipulations analogous to the ones in Section 5.2, we have

$$y(n_1, \ldots, n_k) = \sum_{i = 1}^{K} \alpha_i(n_1, \ldots, n_k)x_i(n_1, \ldots, n_k), \quad (5.31)$$

where

$$x_i(n_1, \ldots, n_k) =$$

$$\sum_{m_1 = 0}^{n_1} \cdots \sum_{m_k = 0}^{n_k} \beta_i(m_1, \ldots, m_k) \cdot u(m_1, \ldots, m_k).$$

$$i = 1, 2, \ldots, K.$$

(5.32)

The expression in (5.32) can be written in matrix form

$$\mathbf{x}_k^i = \mathbf{B}_{i, k}(n_1, \ldots, n_k) \cdot \mathbf{u}_k, \quad i = 1, 2, \ldots, K, \quad (5.33)$$

where $\mathbf{x}_k^i$ and $\mathbf{u}_k$ are vectors of order $([(n_1 + 1) \cdot (n_2 + 1) \ldots (n_k + 1)] \times 1)$ obtained, respectively, by ordering the k-dimensional arrays $\{x_i(i_1, i_2, \ldots, i_k)\}$ and $\{u(i_1, i_2, \ldots, i_k)\}$ $(i_j = 0, 1, \ldots, n_j; j = 1, 2, \ldots, k)$ by means of a lexicographic ordering, described next. The element $u(j_1, j_2, \ldots, j_k)$ occurs in row

$$\sum_{i=1}^{k-1} j_i \prod_{\ell=i}^{k-1} (n_{\ell+1} + 1) + (j_k + 1),$$

for $j_i = 0, 1, \ldots, n_i$, $i = 1, 2, \ldots, k$, of the vector $\mathbf{u}_k$. A similar lexicographical ordering is given to the elements $\{x_i(j_1, j_2, \ldots, j_k)\}$ to form the vector $\mathbf{x}_k^i$, $i = 1, 2, \ldots, K$. $\mathbf{B}_{i,k}(n_1, \ldots, n_k)$ in (5.33) is a square matrix of order $((n_1 + 1) \cdot (n_2 + 1) \ldots (n_k + 1))$, given by (5.34) to (5.36).

The particular structure of the matrices in (5.34) to (5.36) will allow us to perform the summation in (5.32) by means of a $(2^k - 1)$-point recursion. This recursion will be a generalization to k-dimensions ($k > 2$) of the recursion in (5.10) and the justification will be developed next. To do this, the following definitions are, first, introduced.

DEFINITION 5.3. The $\prod_{j=1}^{k} (n_j + 1)$ row-vector $\delta_k^{i_1 i_2 \cdots i_k}$ is defined as

$$\delta_k^{i_1 i_2 \cdots i_k} \triangleq [\mathbf{0}\ \mathbf{0}\ \ldots\ \mathbf{0}\ \delta_{k-1}^{i_2 i_3 \cdots i_k}\ \mathbf{0}\ \ldots\ \mathbf{0}],$$
$$\uparrow \qquad\qquad (5.37)$$
$$(i_1 + 1)\text{-th block position from the right}$$

where $\mathbf{0}$ is a $\prod_{j=2}^{k} (n_j + 1)$ row-vector of zeros, there are n_1 such $\mathbf{0}$ vectors, and the position of the remaining nonzero row-vector is indicated above. Furthermore, the $(n_k + 1)$ row-vector $\delta_1^{i_k}$ is defined as

$$\delta_1^{i_k} \triangleq [0\ 0\ \ldots\ 0\ 1\ 0\ \ldots\ 0].$$
$$\uparrow \qquad\qquad (5.38)$$
$$(i_k + 1)\text{-th position from the right}$$

DEFINITION 5.4. $\sigma_k(j) \triangleq$ sum of all $\binom{k}{j}$ distinct vectors $\delta_j^{i_1 i_2 \cdots i_k}$ with $i_\ell = 0$ or 1 ($\ell = 1, 2, \ldots, k$) and $i_1 + i_2 + \ldots + i_k = j$.

DEFINITION 5.5. $\sigma_{k+1}^i(j) \triangleq$ sum of all $\binom{k}{j}$ distinct vectors $\delta_{k+1}^{i i_2 i_3 \cdots i_{k+1}}$ with $i = 0$ or 1, $i_\ell = 0$ or 1 ($\ell = 2, 3, \ldots, k+1$) and $i_2 + i_3 + \ldots + i_{k+1} = j$.

THEOREM 5.1. *Given the matrices in (5.34) to (5.36), then for any integer $k \geq 1$, we have*

$$\delta_k^{00 \cdots 0} \mathbf{B}_{i,k}(n_1, \ldots, n_k) = \sum_{j=1}^{k} (-1)^{j+1} \sigma_k(j) \mathbf{B}_{i,k}(n_1, \ldots, n_k) + $$
$$+ \delta_k^{00 \cdots 0} \beta_i(n_1, \ldots, n_k). \qquad (5.39)$$

$$\mathbf{B}_{i,k}(n_1, \ldots, n_k) = \begin{bmatrix} \mathbf{B}^0_{i,k-1}(n_2, \ldots, n_k) & \mathbf{B}^1_{i,k-1}(n_2, \ldots, n_k) \\ \mathbf{B}^0_{i,k-1}(n_2, \ldots, n_k) & \mathbf{B}^1_{i,k-1}(n_2, \ldots, n_k) \\ \vdots \\ \mathbf{B}^0_{i,k-1}(n_2, \ldots, n_k) & \mathbf{B}^1_{i,k-1}(n_2, \ldots, n_k) \cdots \mathbf{B}^{n_1}_{i,k-1}(n_2, \ldots, n_k) \end{bmatrix} \tag{5.34}$$

where

$$\mathbf{B}^{i_1 \cdots i_\ell}_{i,k-\ell}(n_{\ell+1}, \ldots, n_k) = \begin{bmatrix} \mathbf{B}^{i_1 \cdots i_\ell 0}_{i,k-\ell-1}(n_{\ell+2}, \ldots, n_k) & \mathbf{B}^{i_1 \cdots i_\ell 1}_{i,k-\ell-1}(n_{\ell+2}, \ldots, n_k) \\ \mathbf{B}^{i_1 \cdots i_\ell 0}_{i,k-\ell-1}(n_{\ell+2}, \ldots, n_k) & \mathbf{B}^{i_1 \cdots i_\ell 1}_{i,k-\ell-1}(n_{\ell+2}, \ldots, n_k) \\ \vdots \\ \mathbf{B}^{i_1 \cdots i_\ell 0}_{i,k-\ell-1}(n_{\ell+2}, \ldots, n_k) & \mathbf{B}^{i_1 \cdots i_\ell 1}_{i,k-\ell-1}(n_{\ell+2}, \ldots, n_k) \cdots \mathbf{B}^{i_1 \cdots i_\ell n_{\ell+1}}_{i,k-\ell-1}(n_{\ell+2}, \ldots, n_k) \end{bmatrix} \tag{5.35}$$

$$\ell = 1, 2, \ldots, k-2, \quad i_j = 0, 1, \ldots, n_j, \quad j = 1, 2, \ldots, \ell$$

$$\mathbf{B}^{i_1 \cdots i_{k-2} r}_{i,1}(n_k) = \begin{bmatrix} \beta_i(i_1, \ldots, i_{k-2}, r, 0) & \beta_i(i_1, \ldots, i_{k-2}, r, 1) \\ \beta_i(i_1, \ldots, i_{k-2}, r, 0) & \beta_i(i_1, \ldots, i_{k-2}, r, 1) \\ \vdots \\ \beta_i(i_1, \ldots, i_{k-2}, r, 0) & \beta_i(i_1, \ldots, i_{k-2}, r, 1) \cdots \beta_i(i_1, \ldots, i_{k-2}, r, n_k) \end{bmatrix} \tag{5.36}$$

$$r = 0, 1, \ldots, n_{k-1}$$

Proof. (by induction)
For $k = 1$

$$\mathbf{B}_{i,\,1}(n_1) = \begin{bmatrix} \beta_i(0) \\ \beta_i(0) & \beta_i(1) \\ \vdots & \vdots \\ \beta_i(0) & \beta_i(1) \ldots \beta_i(n_1) \end{bmatrix}. \qquad (5.40)$$

The last row of $\mathbf{B}_{i,\,1}(n_1)$ in (5.40) can be expressed as

$$[\beta_i(0)\ \beta_i(1)\ \ldots\ \beta_i(n_1)] = [\beta_i(0)\ \beta_i(1)\ \ldots\ \beta_i(n_1-1)0] +$$
$$+\ [0\ 0\ \ldots\ 0\ 1]\beta_i(n_1). \qquad (5.41)$$

Making use of Definitions 5.3 and 5.4, the expression in (5.41) can be written as

$$\delta_1^0\,\mathbf{B}_{i,\,1}(n_1) = \sigma_1(1)\,\mathbf{B}_{i,\,1}(n_1) + \delta_1^0\,\beta_i(n_1), \qquad (5.42)$$

which corresponds to the expression in (5.39) for $k = 1$. From the matrices in (5.34) to (5.36), for any integer $k \geq 1$, we can write

$$[\mathbf{0}\ \mathbf{0}\ \ldots\ \mathbf{0}\ \delta_k^{i_2 i_3\,\cdots\,i_k+1}\mathbf{B}_{i,\,k}^{n_1}(n_2, n_3, \ldots, n_{k+1})] = \qquad (5.43)$$
$$(\delta_{k+1}^{0i_2 i_3\,\cdots\,i_k+1} - \delta_{k+1}^{1i_2 i_3\,\cdots\,i_k+1})\mathbf{B}_{i,\,k+1}(n_1, n_2, \ldots, n_{k+1}).$$

If we now assume that the hypothesis is valid for a fixed integer k, with n_1 fixed, we can write

$$\delta_k^{00\,\cdots\,0}\mathbf{B}_{i,\,k}^{n_1}(n_2, \ldots, n_{k+1}) =$$
$$\sum_{j=1}^{k} (-1)^{j+1}\,\sigma_k(j)\mathbf{B}_{i,\,k}^{n_1}(n_2, \ldots, n_{k+1}) + \qquad (5.44)$$
$$+\ \delta_k^{00\,\cdots\,0}\,\beta_i(n_1, n_2, \ldots, n_{k+1}).$$

Also from (5.43), after setting $i_2 = i_3 = \ldots = i_{k+1} = 0$, we have

$$\delta_{k+1}^{00\,\cdots\,0}\mathbf{B}_{i,\,k+1}(n_1, \ldots, n_{k+1}) =$$
$$\delta_{k+1}^{100\,\cdots\,0}\mathbf{B}_{i,\,k+1}(n_1, \ldots, n_{k+1}) + \qquad (5.45)$$
$$+\ [\mathbf{0}\ \mathbf{0}\ \ldots\ \mathbf{0}\ \delta_k^{00\,\cdots\,0}\mathbf{B}_{i,\,k}^{n_1}(n_2, \ldots, n_{k+1})].$$

Substituting (5.44) in (5.45), we have

$$\delta_{k+1}^{00\cdots0}\mathbf{B}_{i,\,k+1}(n_1,\,\ldots,\,n_{k+1}) =$$

$$\delta_{k+1}^{10\cdots0}\mathbf{B}_{i,\,k+1}(n_1,\,\ldots,\,n_{k+1}) +$$

$$+ \sum_{j=1}^{k} (-1)^{j+1}[\mathbf{0}\ \mathbf{0}\ \ldots\ \sigma_k(j)\mathbf{B}_{i,\,k}^{n1}(n_2,\,\ldots,\,n_{k+1})] +$$

$$+ [\mathbf{0}\ \mathbf{0}\ \ldots\ \mathbf{0}\ \delta_k^{00\cdots0}\beta_{k+1}(n_1,\,\ldots,\,n_{k+1})]. \qquad (5.46)$$

From (5.43), making use of Definitions 5.4 and 5.5, we obtain

$$[\mathbf{0}\ \mathbf{0}\ \ldots\ \mathbf{0}\sigma_k(j)\ \mathbf{B}_{i,\,k}^{n1}(n_2,\,\ldots,\,n_{k+1})] =$$

$$(\sigma_{k+1}^0(j) - \sigma_{k+1}^1(j))\mathbf{B}_{i,\,k+1}(n_1,\,\ldots,\,n_{k+1}). \qquad (5.47)$$

Also, from Definition 5.3, we have

$$[\mathbf{0}\ \mathbf{0}\ \ldots\ \mathbf{0}\ \beta_{k+1}(n_1,\,\ldots\ n_{k+1})] =$$

$$\delta_{k+1}^{00\cdots0}\beta_{k+1}(n_1,\,\ldots\ n_{k+1}). \qquad (5.48)$$

Substituting (5.47) and (5.48) in (5.46), we have

$$\delta_{k+1}^{00\cdots0}\mathbf{B}_{i,\,k+1}(n_1,\,\ldots\ n_{k+1}) = [\delta_{k+1}^{10\cdots0} +$$

$$+ \sum_{j=1}^{k} (-1)^{j+1}(\sigma_{k+1}^0(j) - \sigma_{k+1}^1(j))]\mathbf{B}_{i,\,k+1}(n_1,\,\ldots,\,n_{k+1}) +$$

$$+ \delta_{k+1}^{00\cdots0}\mathbf{B}_{i,\,k+1}(n_1,\,\ldots,\,n_{k+1}). \qquad (5.49)$$

From Definition 5.4, it is straight forward to show that

$$\sigma_{k+1}(j) = \begin{cases} \delta_{k+1}^{10\cdots0} + \sigma_{k+1}^0(1), & j = 1 \\ \sigma_{k+1}^0(j) + \sigma_{k+1}^1(j-1), & j = 2, 3, \ldots, k \\ \sigma_{k+1}^1(k), & j = k+1. \end{cases} \qquad (5.50)$$

Consider now the expression in square brackets in (5.49),

$$\delta_{k+1}^{10\cdots0} + \sum_{j=1}^{k} (-1)^{j+1}(\sigma_{k+1}^0(j) - \sigma_{k+1}^1(j)) = \delta_{k+1}^{100\cdots0} +$$

$$+ \sigma_{k+1}^0(1) + \sum_{j=2}^{k} (-1)^{j+1}(\sigma_{k+1}^0(j) + \sigma_{k+1}^1(j-1)) +$$

$$+ (-1)^{(k+1)+1}\sigma_{k+1}^1(k). \qquad (5.51)$$

Substituting (5.50) in (5.51), we have

$$\delta_{k+1}^{100\,\cdots\,0} + \sum_{j=1}^{k} (-1)^{j+1}(\sigma_{k+1}^{0}(j) - \sigma_{k+1}^{1}(j)) =$$

$$\sum_{j=1}^{k+1} (-1)^{j+1}\sigma_{k+1}(j). \tag{5.52}$$

Substituting (5.52) in (5.49), we have

$$\delta_{k+1}^{00\,\cdots\,0}\mathbf{B}_{i,\,k+1}(n_1,\ldots,n_{k+1}) =$$

$$\sum_{j=1}^{k+1} (-1)^{j+1}\sigma_{k+1}(j)\mathbf{B}_{i,\,k+1}(n_1,\ldots,n_{k+1}) +$$

$$+ \delta_{k+1}^{00\,\cdots\,0}\beta_{k+1}(n_1,\ldots,n_{k+1}), \tag{5.53}$$

which corresponds to the hypothesis for $k+1$. Therefore, by the principle of mathematical induction, the proof is now complete.

The following definition has to be introduced at this point:

DEFINITION 5.6. Define,

$$x_i^j(n_1, n_2, \ldots, n_k) =$$

$$\sum_{s=1}^{\binom{k}{j}} x_i[n_1 - r_s(1), n_2 - r_s(2), \ldots, n_k - r_s(k)]$$

where $\Sigma_{i=1}^{k}r_s(i) = j$, $r_s(i) = 0$ or 1 for each i and the sets $\{r_s(1), r_s(2),$ $\ldots, r_s(k)\}$, $s = 1, 2, \ldots, \binom{k}{j}$ are mutually distinct.

THEOREM 5.2. *Given a k–D, $(k \geq 1)$, positive cone causal LSV system whose impulse response is a K-th order degenerate sequence characterized by expression (5.30). Then, the superposition sum in (5.29) can be implemented by a K-th order state-space model. This state-space model is based on a $(2^k - 1)$-point recursion described by,*

$$x_i(n_1, \ldots, n_k) = \sum_{j=1}^{k} (-1)^{j+1}x_i^j(n_1, \ldots, n_k) +$$

$$\beta_i(n_1, \ldots, n_k)u(n_1, \ldots, n_k)$$

$$i = 1, 2, \ldots, K, \tag{5.54}$$

$$y(n_1, \ldots, n_k) = \sum_{i=1}^{K} \alpha_i(n_1, \ldots, n_k) x_i(n_1, \ldots, n_k). \quad (5.55)$$

Proof. From expressions (5.33) to (5.36), making use of Definition 5.3, we have

$$x_i(n_1 - i_1, n_2 - i_2, \ldots, n_k - i_k) = \delta_k^{i_1 i_2 \cdots i_k} \mathbf{B}_{i,k}(n_1, \ldots, n_k) \mathbf{u}_k$$
$$i = 1, 2, \ldots, K. \quad (5.56)$$

Substituting the expression (5.39) of Theorem 5.1 in (5.56), after setting $i_1 = i_2 = \ldots = i_k = 0$,

$$x_i(n_1, \ldots, n_k) = \sum_{j=1}^{k} (-1)^{j+1} \sigma_k(j) \mathbf{B}_{i,k}(n_1, \ldots, n_k) \mathbf{u}_k +$$
$$+ \delta_k^{00 \cdots 0} \beta_i(n_1, \ldots, n_k) \mathbf{u}_k. \quad (5.57)$$

From (5.56), making use of Definitions 5.4 and 5.6, we have

$$x_i^j(n_1, \ldots, n_k) = \sigma_k(j) \mathbf{B}_{i,k}(n_1, \ldots, n_k) \mathbf{u}_k. \quad (5.58)$$

From the ordering defined for the vector $\mathbf{u}_k$, we have

$$u(n_1, \ldots, n_k) = \delta_k^{00 \cdots 0} \mathbf{u}_k. \quad (5.59)$$

Substituting (5.58) and (5.59) in (5.57), we obtain the expression in (5.54). The recursion is based on ℓ-points, where

$$\ell = \sum_{j=1}^{k} \binom{k}{j} = 2^k - 1. \quad (5.60)$$

The proof is now complete.

The example below serves to clarify the various notations introduced in the preceding discussion. For brevity in exposition, this example tackles the $k = 2$ case, for which the 3-point recursion arrived at in Section 5.2 is verified as a special case of the general result.

EXAMPLE 5.1. Consider (5.32) for the case, $k = 2$. Then, the expression

$$x_i(n_1, n_2) = \sum_{m_1 = 0}^{n_1} \sum_{m_2 = 0}^{n_2} \beta_i(m_1, m_2) u(m_1, m_2),$$

is considered for the case when $n_1 = n_2 = 2$. Then counterpart of the matrix form representation in (5.33) is given below in expanded form. Note that the lexicographical ordering described has been adopted. It is clear that in the $k = 2$ case, this ordering results from a row-by-row scan of each of the arrays $\{x_i(i_1, i_2)\}$ and $\{u(i_1, i_2)\}$.

$$
\begin{bmatrix}
u(0, 0) \\
u(0, 1) \\
u(0, 2) \\
u(1, 0) \\
u(1, 1) \\
u(1, 2) \\
u(2, 0) \\
u(2, 1) \\
u(2, 2)
\end{bmatrix}
=
\begin{bmatrix}
\beta_i(0,0) \\
\beta_i(0,0) & \beta_i(0,1) \\
\beta_i(0,0) & \beta_i(0,1) & \beta_i(0,2) \\
\beta_i(0,0) & & & \beta_i(1,0) \\
\beta_i(0,0) & \beta_i(0,1) & & \beta_i(1,0) & \beta_i(1,1) \\
\beta_i(0,0) & \beta_i(0,1) & \beta_i(0,2) & \beta_i(1,0) & \beta_i(1,1) & \beta_i(1,2) \\
\beta_i(0,0) & & & \beta_i(1,0) & & & \beta_i(2,0) \\
\beta_i(0,0) & \beta_i(0,1) & & \beta_i(1,0) & \beta_i(1,1) & & \beta_i(2,0) & \beta_i(2,1) \\
\beta_i(0,0) & \beta_i(0,1) & \beta_i(0,2) & \beta_i(1,0) & \beta_i(1,1) & \beta_i(1,2) & \beta_i(2,0) & \beta_i(2,1) & \beta_i(2,2)
\end{bmatrix}
\begin{bmatrix}
x_i(0, 0) \\
x_i(0, 1) \\
x_i(0, 2) \\
x_i(1, 0) \\
x_i(1, 1) \\
x_i(1, 2) \\
x_i(2, 0) \\
x_i(2, 1) \\
x_i(2, 2)
\end{bmatrix}
$$

(5.61a)

Specializing the representation in (5.33) to this case, (5.61a) can be written as,

$$\mathbf{x}_2^j = \mathbf{B}_{i,\,2}(n_1,\,n_2)\mathbf{u}_2,$$

where,

$$\mathbf{B}_{i,\,2}(n_1,\,n_2) = \begin{bmatrix} \mathbf{B}_{i,\,1}^0(n_2) \\ \mathbf{B}_{i,\,1}^0(n_2)\ \mathbf{B}_{i,\,1}^1(n_2) \\ \mathbf{B}_{i,\,1}^0(n_2)\ \mathbf{B}_{i,\,1}^1(n_2)\ \mathbf{B}_{i,\,1}^2(n_2) \end{bmatrix}, \tag{5.61b}$$

and

$$\mathbf{B}_{i,\,1}^j(n_2) = \begin{bmatrix} \beta_i(j,\,0) \\ \beta_i(j,\,0)\ \beta_i(j,\,1)\ . \\ \vdots \qquad \vdots \\ \beta_i(j,\,0)\ \beta_i(j,\,1)\ \ldots\ \beta_i(j,\,n_2) \end{bmatrix}. \tag{5.61c}$$

Clearly, (5.61b) and (5.61c) give the relevant specializations of the matrices in (5.34)–(5.36). Applying Definitions 5.3 and 5.4 to the case under consideration, we have

$$\begin{aligned} \boldsymbol{\delta}_2^{00} &= [000 \quad 000 \quad 001], \\ \boldsymbol{\delta}_2^{10} &= [000 \quad 001 \quad 000], \\ \boldsymbol{\delta}_2^{01} &= [000 \quad 000 \quad 010], \\ \boldsymbol{\delta}_2^{11} &= [000 \quad 010 \quad 000], \\ \boldsymbol{\sigma}_2(1) &= \boldsymbol{\delta}_2^{01} + \boldsymbol{\delta}_2^{10}, \\ \boldsymbol{\sigma}_2(2) &= \boldsymbol{\delta}_2^{11}. \end{aligned}$$

Let $\mathbf{r}_j$ denote the j-th row in the matrix $\mathbf{B}_{i,\,2}$ of (5.61a), $j = 1, 2, \ldots, 9$. Clearly

$$\mathbf{r}_9 = \mathbf{r}_8 + \mathbf{r}_6 - \mathbf{r}_5 + [0\ 0\ \ldots\ 0\ \beta_i(2,\,2)]$$

the above equation can be rewritten in the form,

$$\begin{aligned} \boldsymbol{\delta}^{10}\,\mathbf{B}_{i,\,2}(n_1,\,n_2) &= \boldsymbol{\delta}_2^{01}\,\mathbf{B}_{i,\,2}(n_1,\,n_2) + \boldsymbol{\delta}_2^{10}\,\mathbf{B}_{i,\,2}(n_1,\,n_2) - \\ &\quad - \boldsymbol{\delta}_2^{11}\,\mathbf{B}_{i,\,2}(n_1,\,n_2) + \boldsymbol{\delta}_2^{00}\,\beta_i(2,\,2) = \\ &\quad \boldsymbol{\sigma}_2(1)\mathbf{B}_{i,\,2}(n_1,\,n_2) - \boldsymbol{\sigma}_2(2)\mathbf{B}_{i,\,2}(n_1,\,n_2) + \\ &\quad + \boldsymbol{\delta}_2^{00}\,\beta_i(2,\,2), \end{aligned} \tag{5.62}$$

which is the relevant specialization of (5.39). Note that,

$$x_i(n_1 - i,\,n_2 - j) = \boldsymbol{\delta}_2^{ij}\,\mathbf{B}_{i,\,2}(n_1,\,n_2) \cdot \mathbf{u}_2. \tag{5.63}$$

Multiplying (5.62) from the right by $\mathbf{u}_2$ and using (5.63), we get

$$x_i(n_1, n_2) = x_i(n_1, n_2 - 1) + x_i(n_1 - 1, n_2) -$$
$$- x_i(n_1 - 1, n_2 - 1) + \beta_i(n_1, n_2)u(n_1, n_2).$$
$$(5.64)$$

Applying Definition 5.6,

$$x_i^1(n_1, n_2) = x_i(n_1 - 1, n_2) + x_i(n_1, n_2 - 1), \qquad (5.65a)$$

$$x_i^2(n_1, n_2) = x_i(n_1 - 1, n_2 - 1). \qquad (5.65b)$$

Substituting (5.65a) and (5.65b) in (5.64) we get

$$x_i(n_1, n_2) = x_i^1(n_1, n_2) - x_i^2(n_1, n_2) + \beta_i(n_1, n_2)u(n_1, n_2),$$

which is the relevant specialization of (5.54).

5.3.2. *Extension of the k–D State-Space Model to a Causality Hypercone*

The state-space model developed in Section 5.3.1 for a k-dimensional positive cone (or hypercone) causal discrete LSV system, can be naturally extended in order to represent a k-dimensional ($k > 2$) weakly causal discrete LSV system. Definitions 5.7 and 5.8 given next, are introduced in order to reach the desired goal.

DEFINITION 5.7. A k-dimensional causality hypercone C_c; is the intersection of k half hyperplanes $Hp_{i1}, p_{i2}, \ldots, p_{ik}$ ($i = 1, 2, \ldots, k$), where

$$Hp_{i1}, p_{i2}, \ldots, p_{ik} = \{(x_1, \ldots, x_k)|(x_1, \ldots, x_k) \in R^k,$$
$$p_{i1}x_1 + \ldots + p_{ik}x_k \geq 0\}$$
$$(i = 1, 2, \ldots, k)$$

and $p_{ij}(i, j = 1, 2, \ldots, k)$ are non-negative integers, satisfying

$$\det \Phi = 1,$$

where

$$\Phi \triangleq \begin{bmatrix} p_{11} \, p_{12} \cdots p_{1k} \\ \vdots \\ p_{k1} \, p_{k2} \cdots p_{kk} \end{bmatrix}, \qquad (5.66)$$

It is important to note that in a k-dimensional causality cone C_c, any vector $\mathbf{v}$ going from the origin to a point P in C_c can be expressed as a linear combination of the vectors $\mathbf{g}_i$, $i = 1, 2, \ldots, k$. These vectors are called the generator of the causality cone and it can be shown that the generator $\mathbf{g}_i$, $i = 1, 2, \ldots, k$ corresponds to the i-th column of the adjoint of the unimodular transformation matrix Φ.

DEFINITION 5.8. A k–D discrete LSV system with impulse response $h(n_1, \ldots, n_k; m_1, \ldots, m_k)$, is causal on a causality hypercone C_c if and only if

$$h(n_1, \ldots, n_k; m_1, \ldots, m_k) \equiv 0 \quad \text{for} \quad (\ell_1, \ldots, \ell_k) \notin C_c,$$

with $\ell_i = n_i - m_i$ $(i = 1, 2, \ldots, k)$.

The one-to-one and onto mapping Φ, defined in (5.66), maps any integer point in C_c onto a unique integer point in the positive cone Q_1

$$\Phi\colon C_c \cap Z^k \to Q_1 \cap Z^k \quad \text{and}$$

$$\Phi[(0, 0, \ldots, 0)] = (0, 0, \ldots, 0),$$

Given a k–D discrete LSV system with a K-th order degenerate impulse response, which is causal in C_c, then using the mapping Φ defined in (5.66) and its inverse, a state-space model can be derived. The resulting K-th order state-space model will be based on a $(2^k - 1)$-points recursion. For the sake of clarity and brevity in exposition, the procedure is described for the 2–D case, from which the k–D $(k > 2)$ counterpart can be obtained as a direct generalization. Let

$$y(n_1, n_2) = \sum_{(m_1, m_2) \in C_c} \sum h(n_1, n_2; m_1, m_2) u(m_1, m_2),$$

$$(n_1, n_2) \in C_c \quad (5.67)$$

be the input-output relation of a discrete LSV system, and let $\{h(n_1, n_2; m_1, m_2)\}$ be causal in C_c. In addition to this, assume that $\{h(n_1, n_2; m_1, m_2)\}$ is a K-th order degenerate sequence expressed as in (5.2). Using the map Φ in (5.66) with $k = 2$, we map the input, the output and the impulse response in (5.67), as follows

$$\hat{u}(\hat{m}_1, \hat{m}_2) = u(m_1, m_2)\Big|_{\binom{m_1}{m_2} = \Phi^{-1} \cdot \binom{\hat{m}_1}{\hat{m}_2}}, \quad (5.68a)$$

$$\hat{y}(\hat{n}_1, \hat{n}_2) = y(n_1, n_2)\Big|_{\binom{n_1}{n_2} = \Phi^{-1} \cdot \binom{\hat{n}_1}{\hat{n}_2}}, \quad (5.68b)$$

and

$$\hat{h}(\hat{n}_1, \hat{n}_2; \hat{m}_1, \hat{m}_2) = h(n_1, n_2; m_1, m_2)\Big|_{\substack{\binom{m_1}{m_2} = \Phi^{-1}\binom{\hat{m}_1}{\hat{m}_2} \\ \binom{n_1}{n_2} = \Phi^{-1}\binom{\hat{n}_1}{\hat{n}_2}}} \qquad (5.68c)$$

Then, $\hat{h}(\hat{n}_1, \hat{n}_2; \hat{m}_1, \hat{m}_2)$ is first quadrant quarter plane causal. The input array $\hat{u}(\hat{m}_1, \hat{m}_2)$ and the output array $\hat{y}(\hat{n}_1, \hat{n}_2)$ have support on Q_1. In addition to this, from (5.2) and (5.68c) $\hat{h}(\hat{n}_1, \hat{n}_2; \hat{m}_1, \hat{m}_2)$ is a K-th order degenerate sequence, and it can be written as

$$\hat{h}(\hat{n}_1, \hat{n}_2; \hat{m}_1, \hat{m}_2) = \sum_{i=1}^{K} \hat{\alpha}_i(\hat{n}_1, \hat{n}_2)\hat{\beta}_i(\hat{m}_1, \hat{m}_2), \qquad (5.69)$$

where

$$\hat{\alpha}_i(\hat{n}_1, \hat{n}_2) = \alpha_i(n_1, n_2)\Big|_{\binom{n_1}{n_2} = \Phi^{-1}\binom{\hat{n}_1}{\hat{n}_2}}. \qquad (5.70a)$$

$$\hat{\beta}_i(\hat{m}_1, \hat{m}_2) = \beta_i(m_1, m_2)\Big|_{\binom{m_1}{m_2} = \Phi^{-1}\binom{\hat{m}_1}{\hat{m}_2}} \qquad (5.70b)$$

From (5.67) and (5.68), we have

$$\hat{y}(\hat{n}_1, \hat{n}_2) = \sum_{m_1=0}^{\hat{n}_1} \sum_{m_2=0}^{\hat{n}_2} \hat{h}(\hat{n}_1, \hat{n}_2; \hat{m}_1, \hat{m}_2)\hat{u}(\hat{m}_1, \hat{m}_2). \qquad (5.71)$$

For the first quadrant quarter plane causal LSV system in (5.71), with its impulse response given by (5.69), we can now write, using Equations (5.12) to (5.16), a state-space model with support on Q_1.

$$\hat{x}(\hat{n}_1, \hat{n}_2) = \hat{x}(\hat{n}_1 - 1, \hat{n}_2) + \hat{x}(\hat{n}_1, \hat{n}_2 - 1) -$$
$$- \hat{x}(\hat{n}_1 - 1, \hat{n}_2 - 1) + \hat{b}(\hat{n}_1, \hat{n}_2)\hat{u}(\hat{n}_1, \hat{n}_2), \quad (5.72)$$
$$\hat{y}(\hat{n}_1, \hat{n}_2) = \hat{c}(\hat{n}_1, \hat{n}_2)\hat{x}(\hat{n}_1, \hat{n}_2),$$

where

$$\hat{b}(\hat{n}_1, \hat{n}_2) = \begin{bmatrix} \hat{\beta}_1(\hat{n}_1, \hat{n}_2) \\ \vdots \\ \hat{\beta}_K(\hat{n}_1, \hat{n}_2) \end{bmatrix}. \qquad (5.73a)$$

$$\hat{c}(\hat{n}_1, \hat{n}_2) = [\hat{\alpha}_1(\hat{n}_1, \hat{n}_2), \ldots, \hat{\alpha}_K(\hat{n}_1, \hat{n}_2)]. \qquad (5.73b)$$

The initial conditions are:

$$\hat{x}(\hat{n}_1, -1) = 0, \qquad \hat{x}(-1, \hat{n}_2) = 0 \qquad \hat{n}_1, \hat{n}_2 = 0, 1, 2, \ldots$$
$$(5.73c)$$

Mapping back to C_c the state-space equation described in (5.72) and (5.73), (for notational brevity, replace $p_{11}, p_{12}, p_{21}, p_{22}$ in (5.66) by p, r, q, t, respectively)

$$\mathbf{x}(n_1, n_2) = \mathbf{x}(n_1 - t, n_2 + q) + \mathbf{x}(n_1 + r, n_2 - p) -$$
$$- \mathbf{x}(n_1 + r - t, n_2 + q - p) + \mathbf{b}(n_1, n_2)u(n_1, n_2),$$
$$(5.74a)$$

$$y(n_1, n_2) = \mathbf{c}(n_1, n_2)\mathbf{x}(n_1, n_2), \tag{5.74b}$$

where $(n_1, n_2) \in C_c$, and

$$\mathbf{b}(n_1, n_2) = \begin{bmatrix} \beta_1(n_1, n_2) \\ \vdots \\ \beta_K(n_1, n_2) \end{bmatrix}, \tag{5.75a}$$

$$\mathbf{c}(n_1, n_2) = [\alpha_1(n_1, n_2), \ldots, \alpha_K(n_1, n_2)]. \tag{5.75b}$$

The initial conditions are:

$$\mathbf{x}(tn + r, -qn - p) = \mathbf{0}, \qquad \mathbf{x}(-rn - t, pn + q) = \mathbf{0}$$
$$n = 0, 1, \ldots . \tag{5.75c}$$

From the above equations, we can see that the state-space model in (5.74) and (5.75) preserves the three point recursion. It is important to emphasize that in this case the recursion in (5.74a) does not depend on the closest past neighbors of the point currently under consideration.

5.4. STATE-SPACE MODEL FOR THE INVERSE SYSTEM

In Section 5.4.1, the state-space model for the inverse of the system described in (5.10)–(5.17) is first given. In Section 5.4.2, the counterpart in the weakly causal case is considered.

5.4.1 *First Quadrant Quarter-Plane Case*

Consider the state-space model of the first quadrant quarter-plane discrete LSV system described in (5.10)–(5.17). The state-space model of the inverse system, whose state-vector, input and output at the point (n_1, n_2) are, respectively, $\mathbf{z}(n_1, n_2)$, $y(n_1, n_2)$ and $u(n_1, n_2)$ is:

$$\mathbf{z}(n_1, n_2) = \mathbf{A}_I(n_1, n_2)[\mathbf{z}(n_1 - 1, n_2) + \mathbf{z}(n_1, n_2 - 1) -$$
$$- \mathbf{z}(n_1 - 1, n_2 - 1)] + \mathbf{b}_I(n_1, n_2)y(n_1, n_2),$$
$$(5.76a)$$

$$u(n_1, n_2) = c_I(n_1, n_2)[z(n_1 - 1, n_2) + z(n_1, n_2 - 1) -$$
$$- z(n_1 - 1, n_2 - 1)] + d_I(n_1, n_2)y(n_1, n_2),$$

$$(5.76b)$$

with initial conditions

$$z(n_1, -1) = \mathbf{0}, \qquad z(-1, n_2) = \mathbf{0} \quad \text{for} \quad n_i = 0, 1, \ldots, N,$$
$$i = 1, 2, \qquad (5.76c)$$

where

$$A_I(n_1, n_2) = I_K - b(n_1, n_2)[c(n_1, n_2)b(n_1, n_2)]^{-1}c(n_1, n_2),$$
$$(5.77a)$$

and I_K is the K-th order identity matrix,

$$b_I(n_1, n_2) = b(n_1, n_2)[c(n_1, n_2)b(n_1, n_2)]^{-1}, \qquad (5.77b)$$

$$c_I(n_1, n_2) = -[c(n_1, n_2)b(n_1, n_2)]^{-1} c(n_1, n_2), \qquad (5.77c)$$

$$d_I(n_1, n_2) = [c(n_1, n_2)b(n_1, n_2)]^{-1}. \qquad (5.77d)$$

It is important to note at this stage, that the state-space model of the inverse system, given in Equations (5.76) and (5.77), is based also on a three point recursion. Therefore, it will provide a very efficient deconvolution procedure. In addition to this, for a reasonably low order state-space model, there will be no storage problem because the state-vector is only two dimensional and there is no need to store it over the entire input mask.

The state-space model, considered here, for the inverse system exists if and only if

$$c(n_1, n_2)b(n_1, n_2) \neq 0 \quad \text{for} \quad 0 \leq n_i \leq N, \qquad i = 1, 2,$$
$$(5.78)$$

or equivalently, by substituting (5.15) and (5.16) in (5.78) and making use of (5.2), we have that the condition in (5.78) is equivalent to

$$h(n_1, n_2; n_1, n_2) \neq 0 \quad \text{for all} \quad 0 \leq n_i \leq N, \qquad i = 1, 2.$$
$$(5.79)$$

5.4.2. *Weakly Causal Case*

The inverse of the system described in (5.74)–(5.75) is easily shown to be:

$$z(n_1, n_2) = A_I(n_1, n_2)[z(n_1 - t, n_2 + q) + z(n_1 + r, n_2 - p) -$$
$$- z(n_1 + r - t, n_2 + q - p)] +$$
$$+ b_I(n_1, n_2)y(n_1, n_2), \qquad (5.80a)$$

$$u(n_1, n_2) = c_I(n_1, n_2)[z(n_1 - t, n_2 + q) + z(n_1 + r, n_2 - p) -$$
$$- z(n_1 + r - t, n_2 + q - p)] +$$
$$+ d_I(n_1, n_2)y(n_1, n_2), \qquad (5.80b)$$

with initial conditions

$$z(tn + r, -qn - p) = 0, \qquad z(-rn - t, pn + q) = 0,$$
$$n = 0, 1, \ldots \qquad (5.80c)$$

where $A_I(n_1, n_2)$, $b_I(n_1, n_2)$ $c_I(n_1, n_2)$ and $d_I(n_1, n_2)$ have been defined in (5.77).

The necessary and sufficient condition for the existence of the state-space model, in (5.80) is,

$$c_I(n_1, n_2) \cdot b_I(n_1, n_2) \neq 0 \quad \text{for all} \quad (n_1, n_2) \in C_c,$$

or, equivalently,

$$h(n_1, n_2; n_1, n_2) \neq 0 \quad \text{for all} \quad (n_1, n_2) \in C_c,$$

5.5. EXAMPLES OF APPLICATIONS

In this section, two examples are presented which illustrate the application of the state-space model for the inverse system to the restoration of a degraded image, where the degradation is modeled by a LSV system describable by (5.10) to (5.17). In Section 5.5.1, the physical phenomenon responsible for image degradation is motion blur while in Section 5.5.2 the degradation is due to a type of optical aberration, referred to as coma.

5.5.1. *Space-variant Motion Blur*

Consider the one-dimensional motion [5.18]

$$u = g(x; t) = \frac{ax - \alpha t}{t + a} \qquad a, \alpha, u > 0 \tag{5.81}$$
$$t \in [0, T],$$

Using Sawchuk's analysis, this motion is modeled by a LSV system whose impulse response is:

$$h(x, u) = \begin{cases} \dfrac{a}{u + \alpha}, & u \le x \le \dfrac{T + a}{a} u + \dfrac{\alpha T}{a}, \\[2ex] 0 & \text{elsewhere,} \end{cases} \tag{5.82}$$

In order to obtain a 1–D state-space model for this blurring, the impulse response in (5.82) will be approximated in degenerate form. Before carrying out the approximation, it is important to note that the state-space model is causal and, therefore, it will not evaluate the impulse response, $h(x, u)$, at points $x < u$; this will be used to our advantage in extending $h(x, u)$ over these superfluous points. The discontinuity at $x = [(T + a)/a]u + \alpha T/a$, will be slightly smoothed for better results in the approximation. From the above, the function to be approximated, $\hat{h}(x, u)$, is

$$\hat{h}(x, u) = \begin{cases} \dfrac{a}{u + \alpha} & u_1 \le x \le u_2 - D_2, \\[2ex] \dfrac{a}{u + \alpha} \cdot w(x) & u_2 - D_2 \le x \le u_2, \\[2ex] 0 & \text{elsewhere,} \end{cases} \tag{5.83}$$

where

$$u_1 = u - D_1,$$
$$u_2 = \frac{T + a}{a} u + \frac{\alpha T}{a},$$
$$D_2 = (u_2 - u)/10,$$

$w(x)$ is a Hanning window, given as

$$w(x) = .5 \left(1 - \cos\left(\frac{\pi}{D_2} (x - u_2) \right) \right).$$

The constant D_1 will be defined later in this section. The function, $\hat{h}(x, u)$, is plotted, for a fixed but arbitrary value of u, in Figure 5.3.

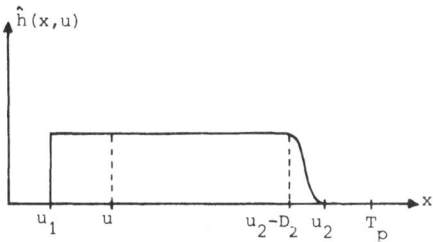

Fig. 5.3. The impulse response of LSV system modeling motion blur.

The u variable will take values over a finite interval, $[0, U]$; then, for any arbitrary but fixed $u \in [0, U]$, we can consider $\hat{h}(x, u)$ to be periodic in x for a sufficiently large period, T_p.

$$T_p > U_2 - U + D_1,$$

where

$$U_2 = \frac{T + a}{a} U + \frac{\alpha T}{a},$$

and D_1 will be defined to be $T_p/4$. Under this assumption $\hat{h}(x, u)$ can be expanded in a Fourier series, which is truncated to yield,

$$h^*(x, u) = a_0(u) + \sum_{k=1}^{K} \left\{ a_k(u)\cos\left(\frac{2\pi}{T_p}kx\right) + \right.$$
$$\left. + b_k(u)\sin\left(\frac{2\pi}{T_p}kx\right) \right\}. \tag{5.84}$$

The $a_k(u)$'s and $b_k(u)$'s in (5.84), obtained in closed form, are

$$a_0(u) = \frac{a}{T(u + \alpha)} (u_2 - u_1 - d_2), \tag{5.85a}$$

$$a_k(u) = \begin{cases} \dfrac{a}{(u + \alpha)} \cdot \{\alpha_k \sin(w_k(u_2 - d)) \cos(w_k d) - \\[2ex] \qquad\qquad - \dfrac{1}{k\pi} \sin(w_k u_1)\}, \qquad \dfrac{1}{D_2} \neq \dfrac{2k}{T}, \\[3ex] \dfrac{a}{(u + \alpha)} \cdot \{\beta_k \sin(w_k(u_2 - d)) \cos(w_k d) - \quad (5.85b) \\[3ex] \qquad - \dfrac{1}{k\pi} \sin(w_k u_1) - \dfrac{d}{T} \cos\left(\dfrac{\pi}{D_2} u_2\right), \\[3ex] \qquad\qquad \dfrac{1}{D_2} = \dfrac{2k}{T} \qquad k = 1, 2, \ldots, \end{cases}$$

$$b_k(u) = \begin{cases} \dfrac{a}{(u + \alpha)} \cdot \{\alpha_k \cos(w_k(u_2 - d_2)) \cos(w_k d) + \\[2ex] \qquad\qquad + \dfrac{1}{k\pi} \cos(w_k u_1)\}, \qquad \dfrac{1}{D_2} \neq \dfrac{2k}{T}, \\[3ex] \dfrac{a}{(u + \alpha)} \cdot \{\beta_k \cos(w_k(u_2 - d)) \cos(w_k d) \dotplus \quad (5.85c) \\[3ex] \qquad + \dfrac{1}{k\pi} \cos(w_k u_1) - \dfrac{d}{T} \sin\left(\dfrac{\pi}{D_2} u_2\right), \\[3ex] \qquad\qquad \dfrac{1}{D_2} = \dfrac{2k}{T} \qquad k = 1, 2, \ldots, \end{cases}$$

where

$$w_k = \frac{2k\pi}{T},$$

$$\alpha_k = 2\left(\frac{w_k}{T(w_k^2 - (\pi/D_2)^2} - \frac{1}{2k\pi}\right),$$

$$\beta_k = 2\left(\frac{1}{2 \cdot T(w_k + \pi/D_2)} - \frac{1}{2k\pi}\right),$$

$$d = D_2/2.$$

In order to simulate the motion blurring phenomenon on a digital computer, the impulse response $\hat{h}(x, u)$ in (5.83) was discretized and $\hat{h}_D(n, m)$ corresponds to the discrete version of (5.83), where

$$\hat{h}_D(n, m) = \hat{h}(n\Delta, m\Delta),$$

for a fixed Δ, such that $U = (N - 1) \cdot \Delta$. The constant U has been previously defined and N corresponds to the size $(N \times N)$ of the input array. The discrete motion blurring is carried out, line by line, by means of the following expression

$$y(n_1, n_2) = \sum_{m_1 = 0}^{n_1} \hat{h}_D(n_1, m_1)u(m_1, n_2). \tag{5.86}$$

$$n_i = 0, 1, \ldots, N - 1, \qquad i = 1, 2.$$

Analogously, before implementing a discrete state-space model for the inverse system, the approximated impulse response $h^*(x, u)$ in (5.84), has to be discretized, and the resulting discrete impulse response $h_D^*(n, m)$ is given by

$$h_D^*(n, m) = h^*(n \cdot \Delta, m \cdot \Delta), \tag{5.87}$$

where Δ has been previously defined. Then, from (5.87) and (5.84)–(5.85), we have

$$h_D^*(n, m) = \sum_{i = 0}^{2 \cdot K} \alpha_i(n)\beta_i(m), \tag{5.88}$$

where

$$\alpha_0(n) = 1, \qquad \beta_0(m) = a_0(m\Delta), \tag{5.89a}$$

$$\alpha_i(n) = \cos\left(n \cdot \frac{2\pi}{T_p} i \cdot \Delta\right), \qquad \beta_i(m) = a_i(m\Delta),$$

$$i = 1, \ldots, K, \tag{5.89b}$$

$$\alpha_i(n) = \sin\left(n \cdot \frac{2\pi}{T_p} i\Delta\right), \qquad \beta_i(m) = b_i(m\Delta),$$

$$i = K + 1, \ldots, 2K, \tag{5.89c}$$

From the $(2 \cdot K + 1)$-th order degenerate sequence, $\{h_D^*(n, m)\}$ derived in (5.88)–(5.89), a state-space model of the inverse system was implemented.

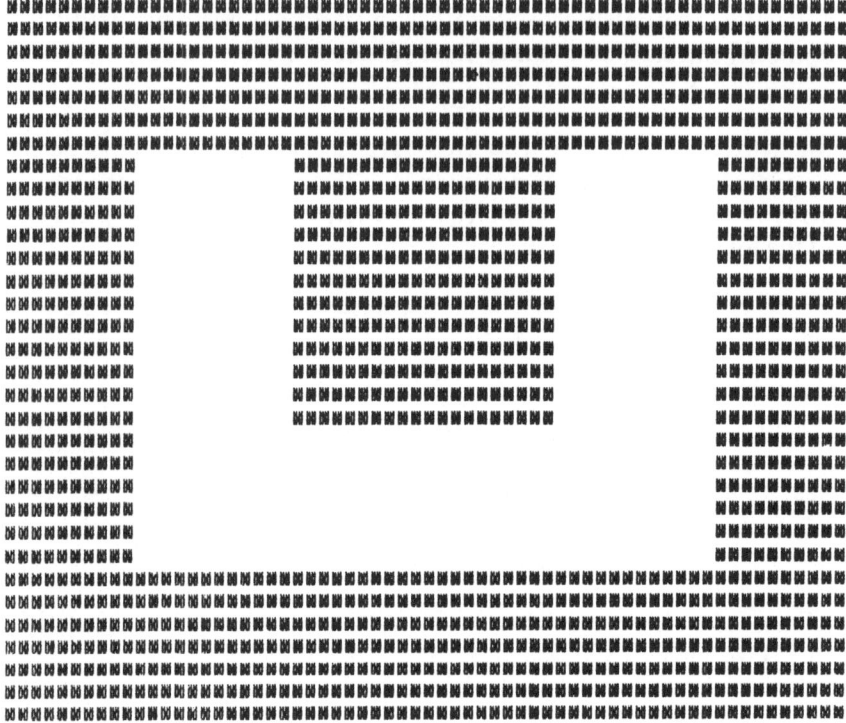

Fig. 5.4a. Original object.

The (32 × 32)-points original object, shown in Figure (5.4a) was
blurred by the simulated motion blur described in (5.86) with the
parameters in (5.82) and (5.84) taken typically as $T = 1$, $\alpha = 2$, $a = 1$
and $\Delta = .2$. The resulting motion blurred image is shown in Figure
(5.4b). From the motion blurred image in Figure (5.4b) by using a
$(2 \cdot K + 1)$-th, $(K = 10)$, state-space model of the inverse system, the
original object was reconstructed. The reconstructed object is shown in
Figure (5.4c). From Figure (5.4c), we can conclude that for a fairly low
order state-space model, the reconstruction is very accurate.

5.5.2. *Coma Aberration*

Within the geometrical optics model of image formation, aberrations are
described by the ray aberration function which is a vector in the image
(output) plane from the Gaussian image point to the point where the ray

actually intersects the image plane. The Gaussian image point is the image in the output plane of a point in the input plane in absence of optical aberrations. Let (r_i, θ_i) and (r_0, θ_0) denote, respectively, the polar coordinates of the image and object planes. It has been shown in a paper by Robbins and Huang [5.19] that the impulse response, $h(r_i, \theta_i; r_0, \theta_0)$, for coma aberration is given by

$$h(r_i, \theta_i; r_0, \theta_0) = \frac{1}{r_0^2} h_0 \left(\frac{r_i \cos(\theta_i - \theta_0) - r_0}{r_0} + 1, \right.$$

$$\left. \frac{r_0 \sin(\theta_i - \theta_0)}{r_0} \right), \qquad (5.90)$$

where the function $h_0(x_i, y_i)$ is the response to an impulse at $r_0 = 1$, $\theta_0 = 0$. The form of $h_0(\cdot)$ for several special cases such as spherical and

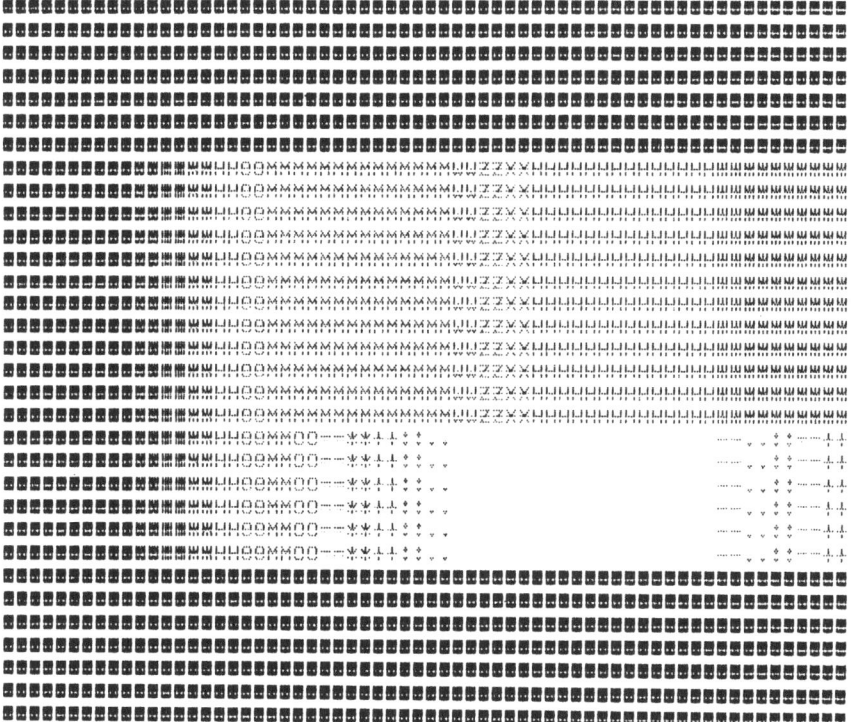

Fig. 5.4b. Motion blurred image.

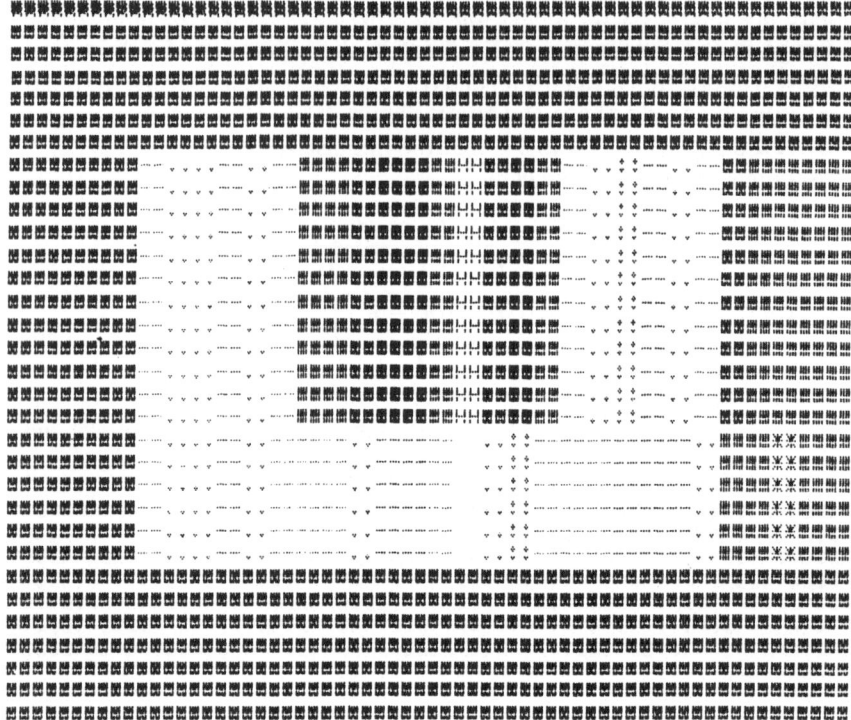

Fig. 5.4c. Restored object from motion blurred image.

coma aberrations are derived in [5.20]. For Cartesian coordinates (x_1, x_2) and (τ_1, τ_2) of the image and object planes, respectively, we have

$$
\begin{aligned}
x_1 &= r_i \cos \theta_i & \tau_1 &= r_0 \cos \theta_0, \\
x_2 &= r_i \sin \theta_i & \tau_2 &= r_0 \sin \theta_0.
\end{aligned}
\tag{5.91}
$$

Substituting (5.91) in (5.90), the impulse response in Cartesian coordinates, $h(x_1, x_2; \tau_1, \tau_2)$, for the coma aberration becomes

$$
h(x_1, x_2; \tau_1, \tau_2) = \frac{1}{\tau_1^2 + \tau_2^2} h_0 \left(\frac{x_1 \tau_1 + x_2 \tau_2}{\tau_1^2 + \tau_2^2}, \frac{x_2 \tau_1 - x_1 \tau_2}{\tau_1^2 + \tau_2^2} \right),
\tag{5.92}
$$

where $h_0(x_1, x_2)$ is the response to an impulse at $\tau_1 = 1$, $\tau_2 = 0$; that is,

$h_0(x_1, x_2) = h(x_1, x_2; 1, 0)$, is given by

$$h_0(x_1, x_2) = \begin{cases} \dfrac{2C}{\sqrt{x_1^2 - 3x_2^2}} & \text{for} \quad (x_1, x_2) \in I, \\[3ex] \dfrac{C}{\sqrt{x_1^2 - 3x_2^2}} & \text{for} \quad (x_1, x_2) \in II, \end{cases} \tag{5.93}$$

where C is a constant and the regions I and II are clearly defined in Figure 5.5.

The impulse response $h(x_1, x_2; \tau_1, \tau_2)$ for the coma aberration in (5.92) is, in general, non-causal, but by imposing some constraint on the radius R_0 in the pattern of the impulse response $h_0(x_1, x_2)$ in (5.93) as shown in Figure 5.5, it is possible to perform the deconvolution, recursively. The

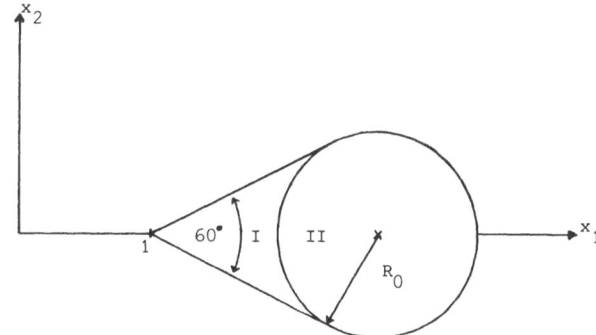

Fig. 5.5. The shape of impulse response of LSV system modeling coma aberration.

restriction on R_0 corresponds to a limitation in the amount of coma aberration to be tolerated, which can be made very small depending on the quality of the lens.

In order to perform the deconvolution, the input plane will be divided into its four quadrants and the deconvolution will be carried out for each one of them. The final reconstructed object is obtained by superimposing the four resulting arrays. For this procedure to be valid, without loss of generality, an object point outside Q_1 should not affect an image point on Q_1. This condition is illustrated in Figure 5.6 and it will be used to determine a bound for R_0.

From Figure 5.6, the inequality $d_v < |\hat{y}|$ is sufficient to guarantee that the object point (τ_1, τ_2) will not affect any image point on Q_1. Simple

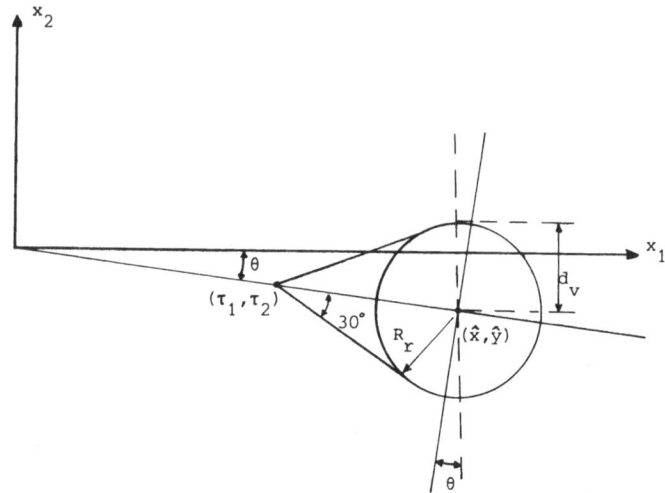

Fig. 5.6. Diagram to determine a maximum value of the radius of the coma pattern.

calculations enable one to get the maximum allowable R_0 required to recursively restore an $(N \times N)$ input array to be

$$R_{0_{max}} = \frac{1}{(\frac{1}{3})^{1/2}(N-1) - 2}$$
(5.94)

As an example, a (31×31)-point original object $u(n_1, n_2)$, $-15 \leq n_i \leq 15$, $i = 1, 2$, which is the counterpart of the (32×32)-point original object in Figure 5.4a was blurred by the come aberration. For this object size, $N = 31$, and $R_{0_{max}} = .065$. The blurred image $y(n_1, n_2)$, $-15 \leq n_i \leq 15$, $i = 1, 2$ was obtained by performing the summation,

$$y(n_1, n_2) = \sum_{m_1 = -15}^{15} \sum_{m_2 = -15}^{15} \hat{h}_D(n_1, n_2; m_1, m_2)u(m_1, m_2) -$$
$$- 15 \leq n_i \leq 15, \qquad i = 1, 2$$

where

$$\hat{h}_D(n_1, n_2; m_1, m_2) = h(n_1\Delta, n_2\Delta; m_1\Delta, m_2\Delta) -$$
$$- 15 \leq n_i \leq 15, -15 \leq m_i \leq 15,$$
$$i = 1, 2$$

and

$$\Delta = 1/15.$$

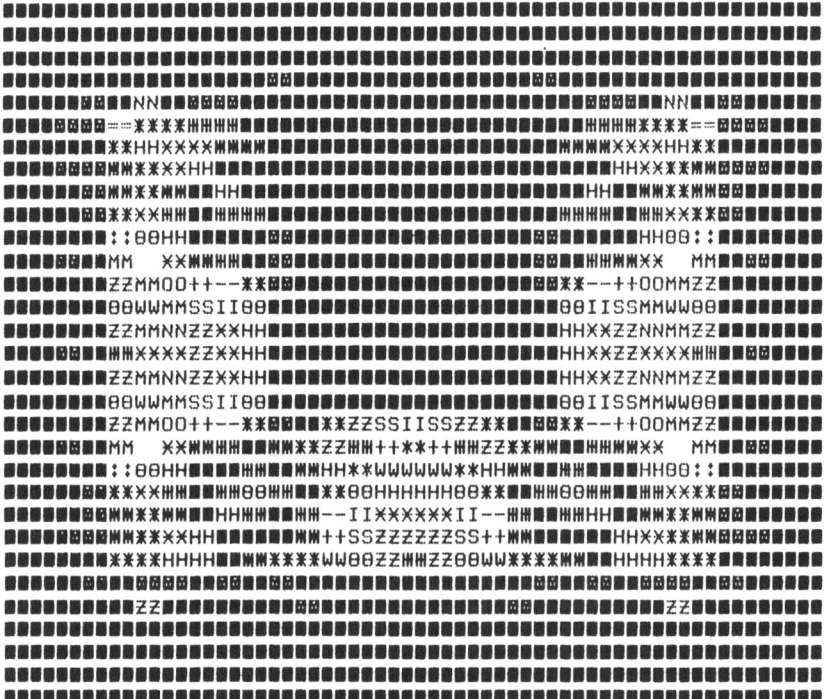

Fig. 5.7. Image blurred by coma aberration.

The resulting blurred image is shown in Figure 5.7. In order to implement the state-space model, the impulse response, $\hat{h}_D(n_1, n_2; m_1, m_2)$, was exactly represented as a degenerate sequence by means of the DFT of $h_D(n_1, n_2; m_1, m_2)$ for each pair (m_1, m_2), $-15 \le m_i \le 15$. This procedure is explained in [5.21]. The symmetry of the impulse response of the coma blurring was used in reducing the amount of DFT's required for this representation. A state-space model of the inverse system was obtained and used in the deconvolution. The deconvoluted image matched exactly the original image.

5.6. CONCLUSIONS

Procedures for compensating for effects which degrade the accuracy of remotely sensed datas by mathematically inverting some of the degrading phenomenon are required in biomedical, industrial, surveillance and earth and space applications. Images to be restored are often degraded by a linear spatially varying operation. Degradations due to motion blurring and optical system distortions often require the imaging systems to be analyzed by modeling the degradation as a linear shift-variant operation. High speed digital computers have been primarily responsible for the development of techniques for image restoration of shift-invariant motion blur, but counterparts of such techniques in the shift-variant case are severely handicapped by the required space-time computational complexity. Attention has been directed to the alleviation of this short-coming and the results arrived at are, first, briefly summarized and then directions for research in the immediate future are provided.

For any 2–D discrete first quadrant quarter-plane causal linear shift-variant (LSV) system, whose impulse response is a K-th order degenerate sequence a K-th order state-space model was obtained. This model is recursive and is based on a three-term recurrence formula relating any point in the state-space model to its three closest past points and the current input. The state-space model was extended in order to model 2–D discrete LSV systems with support on a causality cone. Subsequently, the 2–D quarter-plane causal and weakly causal discrete models were extended to the n–D $(n > 2)$ case. The resulting state-space models are recursive and based on a $(2^n - 1)$-points recurrence formula, which for the causal case used the $(2^n - 1)$-closest neighboring past points in addition to the input in order to compute any current output state. For the weakly causal case, the $(2^n - 1)$ computed points required are not, in general, the closest neighbors to the present output, which is being computed. Conditions for the existence of a 2–D state-space model for the inverse system are readily derivable from the original one. Models for the 2–D LSV system and its inverse can be used to perform analysis and deconvolution problems very efficiently. This can be substantiated from dervided expressions for space-time computational complexities [5.21].

Examples of physically motivated applications making use of the theoretical results developed have been worked out. These applications include effects of 1–D LSV motion blur and the blurring due to Seidel aberrations of a lens; in particular, the 2–D LSV coma aberration was studied in detail. The reconstruction of the original object from the LSV blurred image was carried out successfully by means of the state-space model for the inverse system. For the construction of the LSV model, the impulse responses of the blurring phenomena were approximated in degenerate form via series expansions using orthogonal functions. For details regarding this, see [5.21].

The possibility of using the state-space model of the inverse system for the restoration of the original object from a LSV blurred image in the presence of additive noise is currently under investigation. A causal, discrete counterpart of the integral equation in Phillips [5.22] was implemented, and for various signal-to-noise-ratios (SNR) the deconvolution was performed. It is interesting to note that in this case, there were no large oscillations and as the SNR increased the difference between the original sequence and the deconvolved one decreased. Also the deconvolution of motion blurred and coma blurred images corrupted by noise was performed for different SNR. The restored objects, for a SNR = 100 or larger, were easily recognized in the case of objects with well defined edges, such as a white letter on a black background. The reconstruction of the original object was very poor in the case when the object did not have sharply defined edges.

Since the state-space model developed works very efficiently to deblur images affected by 2–D linear shift-varying blurs, its use, in presence of noise needs to be examined. An obvious approach would be to filter out the additive noise, and, subsequently, obtain, recursively, the restored object using the state-space model of the inverse system, already developed. Specifically, let $x(k_1, k_2)$ be the blurred image with additive noise,

$$x(k_1, k_2) = s(k_1, k_2) + n(k_1, k_2) \qquad (5.95)$$

where $s(k_1, k_2)$ is the spatially-variant blurred image and $n(k_1, k_2)$ is the additive noise. We want to filter out $n(k_1, k_2)$ or in other words we want a satisfactory estimate, $\hat{s}(k_1, k_2)$, of the blurred image. From this obtained estimate, $\hat{s}(k_1, k_2)$, the deconvolution can be implemented, recursively, via the developed state-space model. Some assumptions are necessary before $n(k_1, k_2)$ may be filtered out to satisfaction. A stationary noise, $\{n(k_1, k_2)\}$, which is uncorrelated with $\{s(k_1, k_2)\}$ and having known mean as well as correlation may be assumed. The sequence, $\{s(k_1, k_2)\}$ originates from a spatially blurred object and, therefore, it is inherently non-stationary. We will assume that $\{s(k_1, k_2)\}$ is nonstationary in the mean and also in the autocorrelation. Under these assumptions, we feel that a solution to the filtering problem could be obtained and a possible approach is outlined.

It is possible to transform $\{x(k_1, k_2)\}$ into an approximately stationary process, $\{\bar{x}(k_1, k_2)\}$, given by [5.23],

$$\bar{x}(k_1, k_2) = \bar{s}(k_1, k_2) + \bar{n}(k_1, k_2) \tag{5.96}$$

where $\{\bar{s}(k_1, k_2)\}$ and $\{\bar{n}(k_1, k_2)\}$ are stationary and uncorrelated and the mean and correlation of $\{\bar{n}(k_1, k_2)\}$ are computable from those of $\{n(k_1, k_2)\}$. From the process, $\{\bar{x}(k_1, k_2)\}$, an estimate, $\hat{\bar{s}}(k_1, k_2)$, of $\bar{s}(k_1, k_2)$ is obtained. This is possible to do via use of Wiener filtering theory. Next, an inverse transformation to that employed in order to arrive at (5.96) is applied to $\{\hat{\bar{s}}(k_1, k_2)\}$ and an estimate $\hat{s}(k_1, k_2)$ of the original image $s(k_1, k_2)$ is obtained. It is pointed out that if the noise is nonstationary, further assumptions are necessary (like local stationarity) before a satisfactory solution to the problem is expected.

Linear Restoration of Bilinearly Distorted Image: Several applications require restoration of bilinearly (a special type of nonlinear map) distorted images. Some of these applications and a procedure to restore bilinearly distorted image in the presence of additive noise, by linear filtration is considered in [5.24]. It is also known how 1–D bilinear transformation (shift-invariant or special shift-variant) can be computed by use of 2–D linear optical processors [5.25]. It has also been pointed out in [5.3, p. 219] that properties of n-D bilinear systems can be inferred from the investigation of similar properties in a $2n$-D linear system. These interrelations between a bilinear and a higher dimensional linear system suggest the necessity of investigating into the possibility of restoring bilinearly distorted images using the developed state-space model (of appropriate spatial dimension) for linear shift variant systems.

REFERENCES

[5.1]   H. D'Angelo, *Linear Time-Varying Systems: Analysis and Synthesis*, Allyn and Bacon, Inc. Boston, 1970.
[5.2]   D. G. Tucker, *Circuits with Periodically-Varying Parameters*, Van Nostrand Co., Inc., London, 1964.
[5.3]   N. K. Bose, *Applied Multidimensional Systems Theory*, Van Nostrand Reinhold Co., New York, 1982.
[5.4]   J. W. Goodman, 'Operations Achievable with Coherent Optical Information Processing Systems', *Proc. of IEEE*, January 1977, 65, pp. 29–38.
[5.5]   R. Bamler and J. Hofer-Alfeis, 'Optical Restoration for Rotation Blur by Sequence Deconvolution', Tenth Int. Optical Comp. Conf., M.I.T., Cambridge, MA, April 1983, pp. 146–149.

[5.6]   C. Braccini, 'Scale-invariant, Image Processing by Means of Scaled Transforms or Form-invariant, Linear Shift-variant Filters', *Optics Letters*, Vol. 8, No. 7, July 1983, pp. 392–394.

[5.7]   A. A. Sawchuk, 'Space-variant Image Restoration by Coordinate Transformation', *J. Opt. Soc. Am.*, 64, 1974, pp. 133–144.

[5.8]   B. R. Frieden, 'Restoring with Maximum Likelihood and Maximum Entropy', *J. Opt. Soc. Am.*, 62, April 1972, pp. 511–518.

[5.9]   C. K. Rushforth and R. L. Frost, 'Comparison of some Algorithms for Constructing Space-limited Images', *J. Op. Soc. Am.*, 70, December 1980, pp. 1539–1543.

[5.10] E. S. Angel and A. K. Jain, 'Restoration of Images Degraded by Spatially Varying Point Spread Functions by a Conjugate Gradient Method', *Applied Optics*, 17, July 1978, pp. 2186–2190.

[5.11] S. Kavata and Y. Ichioka, 'Iterative Image Restoration for Linearly Degraded Images 2: Reblurring Procedure', *J. Opt. Soc. Am.*, 70, July 1980, pp. 768–772.

[5.12] S. C. Sahasrabudhe and A. D. Kulkarni, 'Shift Variant Image Degradation and Restoration Using SVD', *Computer Graphics and Image Proc.*, 9, March 1979, pp. 203–212.

[5.13] E. B. Barrett and R. N. Devich, 'Linear Programming Compensation for Space-variant Image Degradation', Abstract in *J. Op. Soc. Am.*, 66, February 1976, p. 172.

[5.14] N. Huang and J. K. Aggarwal, 'On Linear Shift-variant Digital Filters', *IEEE Trans. Circuits & Systems*, 27, 1980, pp. 672–678.

[5.15] L. Zadeh, 'Frequency Analysis of Variable Networks', *Proc. of IRE*, 7, March 1950, pp. 290–299.

[5.16] E. W. Hansen and A. Jablokow, 'State Variable Representation of a Class of Linear Shift-variant systems', *IEEE Transactions on Acoustics, Speech and Signal Processing*, Vol. ASSP-30, No. 6, December 1982, pp. 874–880.

[5.17] A. O. Aboutalib and L. M. Silverman, 'Restoration of Motion Degraded Images', *IEEE Transactions on Circuits and Systems*, Vol. CAS-22, No. 3, March 1975, pp. 278–286.

[5.18] A. A. Sawchuk, 'Space-variant Motion Degradation and Restoration', *Proc. of IEEE*, 60, July 1972, pp. 854–861.

[5.19] G. M. Robbins and T. S. Huang, 'Inverse Filtering for Linear Shift-variant Imaging Systems', *Proceedings of IEEE*, 60, July 1972, pp. 862–872.

[5.20] G. M. Robbins, 'Impulse Response for a Lens with Seidel Aberrations', Res. Lab. Electronics Quarterly Progress Report 93, M.I.T., Cambridge, MA, April 1969.

[5.21] H. M. Valenzuela, 'Modeling of Multidimensional Linear Shift-variant Systems with Applications', Ph.D. Dissertation (based on research conducted under the supervision of N. K. Bose), University of Pittsburgh, Pittsburgh, PA, 1983.

[5.22] D. L. Phillips, 'A technique for the Numerical Solution of Certain Integral Equations of the First Kind', *JACM*, Vol. 9, 1962, pp. 84–97.

[5.23] Rosenfeld, A. (ed.), *Image Modeling*, 'Nonstationary Statistical Image Models and Their Application to Image Data Compression', by B. R. Hunt, Academic Press, New York, 1981.

[5.24] B. E. A. Saleh and W. D. Goeke, 'Linear Restoration of Bilinearly Distorted Images', *J. Op. Soc. Am.*, 70, May 1980, pp. 506–515.

[5.25] B. E. A. Saleh, 'Bilinear Processing of 1–D Signals by Using Linear 2–D Coherent Optical Processors', *Applied Optics*, 21, November 1978, pp. 3408–3411.

Chapter 6

B. Buchberger

Gröbner Bases: An Algorithmic Method in Polynomial Ideal Theory

6.1. INTRODUCTION

Problems connected with ideals generated by finite sets F of multivariate polynomials occur, as mathematical subproblems, in various branches of systems theory, see, for example, [6.1]. The method of Gröbner bases is a technique that provides algorithmic solutions to a variety of such problems, for instance, exact solutions of F viewed as a system of algebraic equations, computations in the residue class ring modulo the ideal generated by F, decision about various properties of the ideal generated by F, polynomial solution of the linear homogeneous equation with coefficients in F, word problems modulo ideals and in commutative semigroups (reversible Petri nets), bijective enumeration of all polynomial ideals over a given coefficient domain etc.

For many years, the work of G. Hermann [6.2] was the only algorithmic method for tackling problems in polynomial ideal theory. Still, her paper is a rich source. However, as pointed out in [6.3] and [6.4], the solution of her main problem "is a multivariate polynomial f in the ideal generated by F?" does not yet give a feasible solution to the "simplification problem modulo an ideal" (i.e. the problem of finding unique representatives in the residue classes modulo the ideal) and to the problem of effectively computing in the residue class ring modulo an ideal.

The method of Gröbner bases, as its central objective, solves the simplification problem for polynomial ideals and, on this basis, gives easy solutions to a large number of other algorithmic problems including Hermann's original membership problem. Also, when compared with Hermann's algorithms, our algorithm that constructs Gröbner bases is of striking simplicity and, depending on the example considered, may get through with intermediate computations using polynomials of relatively low degree. On the other hand, as shown in [6.5] and [6.6], the decision of polynomial ideal congruence intrinsically is a complex problem. In the worst case, therefore, also the method of Gröbner bases may lead to

N. K. Bose (ed.), *Multidimensional Systems Theory*, 184–232

exploding computations. Much work is going on to analyze and predict these phenomena and to extend the applicability of the method.

The method of Gröbner bases was introduced 1965 by this author in [6.7], [6.8] and, starting from 1976, was further refined, generalized, applied and analyzed in a number of papers [6.9]–[6.35]. The basic idea of the method is the transformation of the given set of polynomials F into a certain standard form G, for which in [6.9] the author introduced the name 'Gröbner bases', because Prof. W. Gröbner, the thesis advisor of [6.7] stimulated the research on the subject by asking how a multiplication table for the associative algebra, which is formed by the residue ring modulo a polynomial ideal, can be constructed algorithimically and by presenting a first sketch of an algorithm: He proposed to 'complete' the basis F by adjoining the differences of different representations of power products (modulo the ideal). This, however, is no finite procedure. It was the author's main contribution to see and prove in [6.7], [6.8] that it suffices to adjoin the differences of (the reduced forms of) certain 'critical pairs' (or, equivalently, the reduced form of the 'S-polynomials' [6.7]), which are finite in number.

In retrospect, it seems that the concept of 'Gröbner bases' under the name "standard bases" appeared already one year earlier (1964) in Hironaka's famous paper [6.36]. However, Hironaka only gave an inconstructive existence proof for these bases, whereas in [6.7], together with the concept of such bases, we also presented an algorithm for constructing the bases and only this algorithm allows an algorithmic solution to the various problems shortly mentioned above. An inconstructive existence proof for Gröbner bases may also be found in [6.37]. Hilbert's basis theorem, then, follows as a corollary.

Later (1967) the two basic ideas of our method, critical pairs and completion, where also proposed by Knuth and Bendix [6.38] in the more general context of equations between first order terms. The Knuth-Bendix algorithm now plays an important role in various branches of computer science (abstract data type transformations, equational theorem proving and applications in automated program verification). Recently, the Knuth-Bendix algorithm and the author's own algorithm for constructing Gröbner bases were brought together under a common algorithm structure by R. Llopis de Trias [6.32] and, independently, by P. Le Chenadec [6.39]; see also [6.3] for a general introduction to the "critical-pair completion" algorithm type. On the other hand, the improvements of the author's algorithm were carried over to the Knuth-

Bendix algorithm, see [6.40]. A lot of challenging questions remain to be treated, which, in the near future, might also affect systems theory (for example, decision methods for boolean algebra based on the critical-pair/completion approach, see [6.41].)

In the present paper, a survey on the method of Gröbner bases is given. In Section 6.2, the concept of Gröbner bases is defined and, in Section 6.3, the basic form of the algorithm for constructing Gröbner bases is described. In Section 6.4 an improved version of the algorithm is presented. The improvements are important for the practical feasibility of the computations. In Section 6.5, the algorithm is applied to the simplification problem, the congruence problem and related problems in polynomial ideal theory. In Section 6.6, the algorithm is applied to the exact solution of systems of algebraic equations and related problems. In Section 6.7, it is demonstrated that the S-polynomials have also a significance as the generators of the module of solutions for linear homogeneous equations with polynomial coefficients and an algorithm for a systematic solution of such equations is presented. Gröbner bases for polynomial ideals with integer coefficients are treated in Section 6.8. Some other applications are summarized in Section 6.9. Finally, in Section 6.10, some remarks about specializations, generalizations, implementations and the computational complexity of the algorithm are made.

The emphasis of this paper is on explicit formulation of algorithms (in an easy notation) and on examples. With the exception of some sketches, no proofs of the underlying theorems can be given. However, complete references to the original publications are provided.

6.2. GRÖBNER BASES

Notation

K *a field.*
$K[x_1, \ldots, x_n]$ ring of n-variate polynomials over K.
The following typed variables will be used:
f, g, h, k, p, q polynomials in $K[x_1, \ldots, x_n]$.
F, G finite subsets of $K[x_1, \ldots, x_n]$.
s, t, u power products of the form $x_1^{i_1} \ldots x_n^{i_n}$.
a, b, c, d elements in K.
i, j, l, m natural numbers.
Let $F = \{f_1, \ldots, f_m\}$. By 'Ideal(F)' we will denote "the ideal generated

by F" (i.e. the set

$$\{ \sum_{1 \leqslant i \leqslant m} h_i \cdot f_i | h_i \epsilon K[x_1, \ldots, x_n] (i = 1, \ldots, m)\}).$$

Furthermore, we will write '$f \equiv_F g$' for "f is congruent to g modulo Ideal(F)" (i.e. $f\text{-}g\epsilon$Ideal(F)).

Before one can define the notion of Gröbner bases the notion of 'reduction' must be introduced. For this it is necessary to fix a total odering $<_T$ of the power products $x_1^{i_1} \ldots x_n^{i_n}$, for example, the 'total degree ordering' (which is $1 <_T x <_T y <_T x^2 <_T xy <_T y^2 <_T x^3 <_T x^2 y <_T xy^2 <_T y^3 <_T \ldots$ in the case of two variables) or the 'purely lexicographical ordering' (which is $1 <_T x <_T x^2 <_T x^3 <_T \ldots y <_T xy <_T x^2 y <_T \ldots <_T y^2 <_T xy^2 <_T \ldots$ in the case of two variables). In fact, any total ordering is suitable, which at least has the following two properties:

(T1) $\qquad 1 <_T t \quad$ *for all* $\quad t \neq 1$,

(T2) $\qquad$ if $s <_T t \quad$ *then* $\quad s \cdot u <_T t \cdot u$.

A total ordering satisfying (T1) and (T2) will be called 'admissible'. For the sequel, assume that an arbitrary $<_T$ has been fixed. With respect to the chosen $<_T$, we use the following notation.

Notation

Coefficient(g, t) $\qquad$ the coefficient of t in g.

LeadingPowerProduct(f) $\qquad$ the maximal power product ($w.r.t.$ $<_T$) occurring with non-zero coefficient in f.

LeadingCoefficient(f) $\qquad$ the coefficient of the LeadingPower-Product(f).

DEFINITION 6.1 [6.7], [6.8].

$g \rightarrow_F h$ (read: 'g *reduces* to h modulo F') iff there exists $f\epsilon F$, b and u such that

$$g \rightarrow_{f, b, u} \quad \text{and} \quad h = g - b \cdot u \cdot f.$$

$g \rightarrow_{f, b, u}$ (read: 'g is reducible using f, b, u') iff Coefficient(g, $u \cdot$ LeadingPowerProduct(f)) $\neq 0$, $b =$ Coefficient(g, $u \cdot$ LeadingPower-Product(f))/LeadingCoefficient(f) $\qquad \bullet$

Hence, roughly, g reduces to h modulo F iff a monomial in g can be deleted by the subtraction of an appropriate multiple $b \cdot u \cdot f$ of a polynomial f in F yielding h. Thus, the reduction may be viewed as one step in a generalized division.

EXAMPLE 6.1. Consider $F: = \{f_1, f_2, f_3\}$, where

$$f_1: = 3x^2y + 2xy + y + 9x^2 + 5x - 3,$$
$$f_2: = 2x^3y - xy - y + 6x^3 - 2x^2 - 3x + 3,$$
$$f_3: = x^3y + x^2y + 3x^3 + 2x^2.$$

The polynomials f_1, f_2, f_3 are ordered according to the purely lexicographical ordering. The leading power products are x^2y, x^3y, x^3y, respectively, and the leading coefficients are 3, 2, and 1. Consider

$$g: = 5y^2 + 2x^2y + 5/2xy + 3/2y + 8x^2 + 3/2x - 9/2.$$

Modulo F, g reduces, for example, to

$$h: = 5y^2 + 7/6xy + 5/6y + 2x^2 - 11/6x - 5/2.$$

Namely,

$$g \rightarrow_{f, b, u} \quad \text{for} \quad f: = f_1, \qquad b: = 2/3, \qquad u: = 1$$

because Coefficient$(g, 1 \cdot x^2y) = 2 \neq 0$ and $b = $ Coefficient$(g, 1 \cdot x^2y)/$ LeadingCoefficient(f_1), and

$$h = g - (2/3) \cdot 1 \cdot f_1.$$

DEFINITION 6.2.
h is in *normal form* (or reduced form) modulo F iff there is no h' such that $h \rightarrow_F h'$.
h is *a normal form of g* modulo F iff there is a sequence of reductions

$$g = k_0 \rightarrow_F k_1 \rightarrow_F k_2 \rightarrow_F \ldots \rightarrow_F k_m = h$$

and h is in normal form modulo F.
An algorithm S is called a *normal form algorithm* (or simplifier) iff for all F and g:

$$S(F, g) \text{ is a normal form of } g \text{ modulo } F.$$

LEMMA 6.1 [6.7] [6.9].
The following algorithm is a normal form algorithm:

ALGORITHM 6.1 ($h: = $ NormalForm(F, g)).

$h: = g$

while exist $f \epsilon F$, b, u such that $h \rightarrow_{f, b, u}$ *do* choose $f \epsilon R$, b, u such that $h \rightarrow_{f, b, u}$ and $u \cdot$ LeadingPowerProduct(f) is maximal (w.r.t. $<_T$)

$h: = h - b \cdot u \cdot f$ ●

The correctness of this algorithm should be clear. For the correctness, the selection of the maximal product $u \cdot$ LeadingPowerProduct(f) is not mandatory. However, this choice is of crucial importance for efficiency. The termination of the algorithm is guaranteed by the following lemma.

LEMMA 6.2 [6.7], [6.9]. For all F: $\rightarrow_F$ is a noetherian relation (i.e. there is no infinite sequence $k_0 \rightarrow_F k_1 \rightarrow_F k_2 \rightarrow_F \ldots$).

EXAMPLE 6.2. h in the Example 6.1 is in normal form modulo F: no power product occurring in h is a multiple of the leading power product of one of the polynomials in F. Thus, no reduction is possible. Another example:

$$x^3y \rightarrow_{f_1} - 2/3x^2y - 1/3xy - 3x^3 - 5/3x^2 + x = :g_1.$$

g_1 can be further reduced:

$$g_1 \rightarrow_{f_1} 1/9xy + 2/9y - 3x^3 + 1/3x^2 + 19/9x - 2/3 = :g_1'.$$

g_1' is in normal form modulo F. g_1', hence, is a normal form of x^3y modulo F. Actually, g_1' may be the result of applying the algorithm 'NormalForm' to x^3y (depending on how the instruction 'choose $f \epsilon F$, such that . . .' in the algorithm is implemented). In this example, a second reduction is possible:

$$x^3y \rightarrow_{f_2} 1/2xy + 1/2y - 3x^3 + x^2 + 3/2x - 3/2 = :g_2.$$

g_2 is already in normal form modulo F.

From the example one sees that, in general, it is possible that, modulo F, g_1 and g_2 are normal forms of a polynomial g, but $g_1 = g_2$. Those sets F, for which such a situation does not occur, play the crucial role for our approach to an algorithmic solution of problems in polynomial ideal theory:

DEFINITION 6.3 [6.7], [6.9]. F is called a *Gröbner basis* (or Gröbner set) iff
for all g, h_1, h_2:
   if h_1 and h_2 are normal forms of g modulo F then $h_1 = h_2$.   ●
It is the central theme of this paper to show that
   (a) for those sets F that are Gröbner bases, a number of important
algorithmic problems (that are formulated in terms of Ideal(F)) can be
solved elegantly and
   (b) those sets F, which are not Gröbner bases, can be transformed into
sets G, that are Gröbner bases and generate the same ideal.
   Most of the algorithmic applications of Gröbner bases are based on the
following fundamental property of Gröbner bases.

THEOREM 6.1 [6.7], [6.9], [6.22] (Characterization Theorem for Gröbner
bases). Let S be an arbitrary normal form algorithm. The following
properties are equivalent:
(GB1) F is a Gröbner basis.
(GB2) For all f, g: $f \equiv_F g$ iff $S(F, f) = S(F, g)$.   ●
   (GB1) is also equivalent to:
(GB3) $\to_F$ has the 'Church-Rosser' property.
(GB3) links Gröbner bases with analogous concepts for equations of first
order terms and the Knuth-Bendix algorithm. For details see [6.3].
(GB3) is not needed in this paper. The following lemma is helpful in
establishing this link.

LEMMA 6.3 [6.22], [6.30] (Connection between reduction and congruence): For all F, f, g:

$$f \equiv_F g \quad \text{iff} \quad f \leftrightarrow_F^* g.$$

(Here, $\leftrightarrow_F^*$ is the reflexive, symmetric, transitive closure of $\to_F$, i.e.

$$f \leftrightarrow_F^* g \quad \text{iff} \quad \text{there exists a sequence}$$
$$f = k_0 \leftrightarrow_F k_1 \leftrightarrow_F k_2 \leftrightarrow_F \ldots \leftrightarrow_F k_m = g,$$

where

$$f \leftrightarrow_F g \quad iff \quad (f \to_F g \quad \text{or} \quad g \to_F f)).$$ ●

(GB2) immediately shows that, for Gröbner bases F, the decision
problem '$f \equiv_F g$' is algorithmically decidable (uniformly in F). For
Gröbner bases, other computability problems will have similarly easy
solutions: see Sections 5–9.

6.3. ALGORITHMIC CONSTRUCTION OF GRÖBNER BASES

Before we give the algorithmic applications of Gröbner bases we show how it may be decided whether a given set F is a Gröbner basis and how Gröbner bases may be constructed. For this the notion of an 'S-polynomial' is fundamental:

DEFINITION 6.4 [6.7], [6.8], [6.9].
The 'S-polynomial corresponding to f_1, f_2' is
$S\text{Polynomial}(f_1, f_2)$: $= u_1 \cdot f_1 - (c_1/c_2) \cdot u_2 \cdot f_2$,
where $c_i = \text{LeadingCoefficient}(f_i)$,
u_i is such that $s_i \cdot u_i =$ the least common multiple of s_1, s_2 and
$s_i = \text{LeadingPowerProduct}(f_i)$ $(i = 1, 2)$.

EXAMPLE 6.3. For f_1, f_2 as in Example 6.1, the $S\text{Polynomial}(f_1, f_2)$ is

$$2x^2y + 5/2xy + 3/2y + 8x^2 + 3/2x - 9/2. \qquad \bullet$$

Note that the least common multiple of s_1 and s_2 is the minimal power product that is reducible both modulo f_1 and modulo f_2. The algorithmic criterion for Gröbner bases is formulated in the following theorem, which forms the core of the method:

THEOREM 6.2 (Buchberger [6.7], [6.8], [6.9], [6.22]; Algorithmic Characterization of Gröbner bases). Let S be an arbitrary normal form algorithm. The following properties are equivalent:
(GB1) F is a Gröbner basis.
(GB4) For all $f_1, f_2 \epsilon F$: $S(F, S\text{Polynomial}(f_1, f_2)) = 0$. $\bullet$

(GB4), indeed, is a decision algorithm for the property 'F is a Gröbner basis': one only has to consider the finitely many pairs f_1, f_2 of polynomials in F, compute the corresponding S-polynomials and see whether they reduce to zero by application of the normal form algorithm S. In addition, Theorem 6.2 is the basis for the central Algorithm 6.2 of this paper for solving the following problem.

PROBLEM 6.1.
Given F.
Find G, such that $\text{Ideal}(F) = \text{Ideal}(G)$ and G is a Gröbner basis.

ALGORITHM 6.2 (Buchberger [6.7], [6.8]) for Problem 6.1.

G: $= F$

B: $= \{\{f_1, f_2\}|f_1, f_2 \in G, f_1 \neq f_2\}$

while $B \neq \emptyset$ *do*

    $\{f_1, f_2\}$: $= a$ pair in B

    B : $= B - \{\{f_1, f_2\}\}$

    h : $=$ SPolynomial(f_1, f_2)

    h' : $=$ NormalForm(G, h)

    if $h' \neq 0$ *then*

        B: $= B \cup /\{\{g, h'\}|g \in G\}$

        G: $= G \cup /\{h'\}$.   •

The partial correctness of this algorithm, essentially, relies on Theorem 6.2. The termination can be shown in two ways, see [6.8], [6.17]. (Sketch of the first method [6.17]: One considers the sequence of ideals Ideal$(P_1) \subset$ Ideal$(P_2) \subset \ldots$, where P_i is the set of leading power products of polynomials in G_i and G_i is the value of G after G has been extended for the i-th time. It is easy to see, that the inclusions in this sequence are proper. Hence, by Hilbert's theorem on ascending chains of ideals in $K[x_1, \ldots, x_n]$, see [6.42], the sequence must be finite. Sketch of the second method [6.8]: One uses Dickson's lemma [6.43], which, applied to the present situation, shows that a sequence $t_1, t_2, \ldots$ of power products with the property that, for all j, t_j is not a multiple of any of its predecessors, must be finite. Actually, if t_i is the leading power product of the i-th polynomial adjoined to G in the course of the algorithm ($i = 1, 2, \ldots$), then the sequence $t_1, t_2, \ldots$ has this property and, hence, must be finite. This is the way, the termination of the algorithm was first proven in [6.8], where Dickson's lemma was reinvented. Hilbert's basis theorem can be obtained as a corollary in this approach, see [6.37].)

EXAMPLE 6.4. Starting from the set F of Example 6.1, we first choose, for instance, the pair f_1, f_2 and calculate

    SPolynomial$(f_1, f_2) =$
    $2x^2y + 5/2xy + 3/2y + 8x^2 + 3/2x - 9/2$.

Reduction of this polynomial to a reduced form yields

$$7/6xy + 5/6y + 2x^2 - 11/6x - 5/2.$$

We adjoin this polynomial to G in the form

$$f_4: = xy + 5/7y + 12/7x^2 - 11/7x - 15/7,$$

where we normalized the leading coefficient to 1. (This normalization is not mandatory. However, as a matter of computational experience, it may result in drastic savings in computations over the rationals. Theoretically, this phenomenon is not yet well understood. Investigations of the kind done for Euclid's algorithm should be worthwhile, see [6.44] for a survey on these questions.)

Now we choose, for example, the pair f_1 and f_4:

$$SPolynomial(f_1, f_4) = 1 \cdot f_1 - (3/1) \cdot x \cdot f_4 =$$
$$-1/7xy + y - 36/7x^3 + 96/7x^2 + 80/7x - 3.$$

Reduction of this polynomial, by subtraction of $-(1/7) \cdot f_4$ (and normalization), yields the new polynomial.

$$f_5: = y - 14/3x^3 + 38/3x^2 + 61/6x - 3.$$

Furthermore, $SPolynomial(f_4, f_5) = 1 \cdot f_4 - (1/1) \cdot x \cdot f_5$. By subtracting $(5/7) \cdot f_5$ and normalization we obtain

$$f_6: = x^4 - 2x^3 - 15/4x^2 - 5/4x.$$

Finally, the reduction of $SPolynomial(f_1, f_3) = x \cdot f_1 - (3/1) \cdot 1 \cdot f_3$ leads to

$$f_7: = x^3 - 5/2x^2 - 5/2x.$$

The reduction of the S-polynomials of all the remaining pairs yields zero and, hence, no further polynomials need to be adjoined to the basis. For example,

$$SPolynomial(f_6, f_7) = 1/2x^3 - 5/4x^2 - 5/4x$$

reduces to zero by subtraction of $1/2 f_7$. Hence, a Gröbner basis corresponding to F is

$$G: = \{f_1, \ldots, f_7\}.$$

DEFINITION 6.5 [6.10]. F is a *reduced Gröbner basis* iff F is a Gröbner basis and for all $f \in F$: f is in normal form modulo $F - \{f\}$ and LeadingCoefficient$(f) = 1$.

EXAMPLE 6.5. G in Example 6.4 is not a reduced Gröbner basis: For example, f_1 reduces to zero modulo $\{f_2, \ldots, f_7\}$. By successively reducing all polynomials of a Gröbner basis modulo all the other polynomials in the basis and normalizing the leading coefficients to 1, one always can transform a Gröbner basis into a reduced Gröbner basis for the same ideal. We do not give a formal description of this procedure, because it will be automatically included in the improved version of the algorithm below. In the example, also f_2, f_3, f_4, and f_6 reduced to zero and f_5 reduces to

$$f_5': = y + x^2 - 3/2x - 3.$$

Hence, the reduced Gröbner basis corresponding to F is

$$G': = \{f_5', f_7\} = \{y + x^2 - 3/2x - 3, x^3 - 5/2x^2 - 5/2x\}.$$

THEOREM 6.3 (Buchberger [6.10]: Uniqueness of reduced Gröbner bases). If $\text{Ideal}(F) = \text{Ideal}(F')$ and F and F' are both reduced Gröbner bases then $F = F'$.

DEFINITION 6.6. Let GB be the function that associates with every F a G such that $\text{Ideal}(F) = \text{Ideal}(G)$ and G is a reduced Gröbner basis.   •

By what was formulated in Theorems 6.2, 6.3, Algorithm 6.2 and the remarks in Example 6.5 we, finally, obtain the following main theorem, which summarizes the basic algorithmic knowledge about Gröbner bases.

MAIN THEOREM 6.4 (Buchberger 1965, 1970, 1976).
GB is an algorithmic function that satisfies for all F, G:
(SGB1) $\text{Ideal}(F) = \text{Ideal}(GB(F))$,
(SGB2) if $\text{Ideal}(F) = \text{Ideal}(G)$ then $GB(F) = GB(G)$,
(SGB3) $GB(F)$ is a reduced Gröbner basis.

6.4. AN IMPROVED VERSION OF THE ALGORITHM

For the tractability of practical examples it is crucial to improve the algorithm. There are three main possibilities for achieving a computational speed-up:
(1) The order of selection of pairs $\{f_1, f_2\}$ for which the S-polynomials are formed, though logically insignificant, has a crucial influence on the

complexity of the algorithm. As a general rule, pairs whose least common multiple of the leading power products is minimal with respect to the ordering $<_T$ should be treated first. This, in connection with (2), may drastically reduce the computation time.

(2) Each time a new polynomial is adjoined to the basis, all the other polynomials may be reduced using also the new polynomial. Thereby, many polynomials in G may be deleted again. Such reductions may initiate a whole cascade of reductions and cancellations. Also, if this procedure is carried out systematically in the course of the algorithm, the final result of the algorithm automatically is a *reduced* Gröbner basis. The reduction of the polynomials modulo the other polynomials in the basis should also be performed at the beginning of the algorithm.

(3) Whereas (1) and (2) are strategies that need no new theoretical foundation, the following approach is based on a refined theoretical result [6.19], which has proven useful also in the general context of 'critical-pair/completion' algorithms, in particular for the Knuth-Bendix algorithm: The most expensive operations in the algorithm are the reductions of the h' modulo G in the *while*-loop. We developed a 'criterion' that, roughly, allows to detect that certain S-polynomials h can be reduced to zero, without actually carrying out the reduction. This can result in drastic savings. Using this criterion, in favourable situations, only $0(l)$ S-polynomials must be considered instead of $0(l^2)$, where l is the number of polynomials in the basis. (Of course, in general, l is dynamically changing and, therefore, the effect of the criterion is very hard to assess, theoretically).

Strategy 1. was already used in [6.7], [6.8]. Also, the correctness of the reduction and cancellation technique sketched in (2) was already shown in [6.7], [6.8]. The criterion described in (3) was introduced and proven correct in [6.19], details of the correctness proof may be found in [6.20].

Before we give the details of the improved version of the algorithm based on (1)–(3) we present a rough sketch:

In addition to G and B, we use two sets R and P. R contains polynomials of G, which can be reduced modulo the other polynomials of G. As long as R is non-empty, we reduce the polynomials in R and store the resulting reduced polynomials in P. Only when R is empty, we adjoin the reduced polynomials in P to G and determine the new pairs in B for which the S-polynomials have to be considered. If an S-polynomial for a pair in B is reduced with a non-zero result h', h' is put into P and, again, polynomials in G are sought that are reducible with respect to h'. Such polynomials are put into R and we continue with the systematic reduction of R. We now give the details.

PROBLEM 6.2.
Given: F.
Find: G, such that $\text{Ideal}(F) = \text{Ideal}(G)$ and G is a reduced Gröbner basis.

ALGORITHM 6.3 (Buchberger [6.19]) for Problem 6.2.

$R: = F; P: = \emptyset; G: = \emptyset; B: = \emptyset$

Reduce All (R, P, G, B); New Basis (P, G, B)

while $B \neq \emptyset$ do

$\quad\quad \{f_1, f_2\}: = $ a pair in B whose $LCM(LP(f_1), LP(f_2))$ is minimal
$\quad\quad\quad\quad$ w.r.t. $<_T$

$\quad\quad B: = B - \{\{f_1, f_2\}\}$

$\quad\quad$ *if (not* Criterion1(f_1, f_2, G, B) *and*

$\quad\quad\quad\quad$ *not* Criterion2(f_1, f_2)*) then*

$\quad\quad\quad\quad\quad h: = $ NormalForm$(G, S$Polynomial$(f_1, f_2))$

$\quad\quad\quad\quad\quad$ *if $h \neq 0$ then*

$\quad\quad\quad\quad\quad\quad\quad G_0: = \{g \epsilon G | LP(h) \leq_M LP(g)\}$

$\quad\quad\quad\quad\quad\quad\quad R: = G_0; P: = \{h\}; G: = G - G_0$

$\quad\quad\quad\quad\quad\quad\quad B: = B - \{\{f_1, f_2\} | f_1 \epsilon G_0 \text{ or } f_2 \epsilon G_0\}$

$\quad\quad\quad\quad\quad\quad\quad$ ReduceAll(R, P, G, B); NewBasis(P, G, B).

Subalgorithm Reduce All *(transient : R, P, G, B):*

while $R \neq \emptyset$ do

$\quad\quad h: = $ an element in R; $R: = R - \{h\}$;

$\quad\quad h: = $ NormalForm$(G \cup P, h)$

$\quad\quad$ *if $h \neq 0$ then*

$\quad\quad\quad\quad G_0: = \{g \epsilon G | LP(h) \leq_M LP(g)\}$

$\quad\quad\quad\quad P_0: = \{p \epsilon P | LP(h) \leq_M LP(p)\}$

$$G : = G - G_0$$
$$P : = P - P_0$$
$$R : = R \cup G_0 \cup P_0$$
$$B : = B - \{\{f_1, f_2\} \epsilon B | f_1 \epsilon G_0 \quad \text{or} \quad f_2 \epsilon G_0\}$$
$$P : = P \cup \{h\}.$$

Subalgorithm New Basis (*transient* : P, G, B):

$$G: = G \cup P$$
$$B: = B \cup \{\{g, p\} | g \epsilon G, p \epsilon P, g \neq p\}$$
$$H: = G; K: = \emptyset$$

while $H \neq \emptyset$ *do*

$\quad$ $h: =$ an element in $H; H: = H - \{h\}$

$\quad$ $k: = \text{NormalForm}(G - \{h\}, h); K: = K \cup \{k\}$

$G: = K.$

Subalgorithm Criterion1(f_1, f_2, G, B): $\Leftrightarrow$ there exists a $p \epsilon G$ such that

$$f_1 \neq p, p \neq f_2,$$
$$LP(P) \leq_M LCM(LP(f_1), LP(f_2)),$$
$$\{f_1, p\} \text{ not in } B \text{ and } \{p, f_2\} \text{ not in } B.$$

Subalgorithm Criterion2(f_1, f_2): $\Leftrightarrow$

$$LCM(LP(f_1), LP(f_2)) = LP(f_1) \cdot LP(f_2).$$

Abbreviations

$LP(g)$ $\quad$ the leading power product of g,

$LCM(s, t)$ $\quad$ the least common multiple of s and t,

$s \leq_M t$ $\quad$ t is a multiple of s. $\qquad\qquad\qquad\qquad\qquad$ •

The correctness of this improved version of the algorithm is based on the following lemma and theorem.

LEMMA 6.4 [6.7], [6.8]. For arbitrary F, f_1, f_2:
If $LP(f_1) \cdot LP(f_2) = LCM(LP(f_1), LP(f_2))$, then $SPolynomial$ (f_1, f_2) can always be reduced to zero modulo F.

THEOREM 6.5 (Buchberger 1979 [6.19]; detection of unnecessary reductions of S-polynomials). Let S be an arbitrary normal form algorithm. The following properties are equivalent:
(GB1) F is a Gröbner basis.
(GB5) For all f, $g \epsilon F$ there exist h_1, h_2, . . ., $h_k \epsilon F$ such that

$$f = h_1, \qquad g = h_k,$$
$$LCM(LP(h_1), \ldots, LP(h_k)) \leq_M LCM(LP(f), LP(g)),$$
$$S(F, \text{SPolynomial}(h_i, h_{i+1})) = 0 \ (\text{for } 1 \leq i < k). \qquad \bullet$$

Lemma 6.4 guarantees that we need not consider the S-polynomial of two polynomials f_1 and f_2, whose leading power products satisfy the condition stated in the lemma (Criterion2). The iteration of Criterion1 in Algorithm 6.3 guarantees that, upon termination of the algorithm, condition (GB5) is satisfied for G and, hence, G is a Gröbner basis.

EXAMPLE 6.6. Let $F: = \{f_1, f_2, f_3\}$, where

$$f_1: = x^3yz - xz^2, \qquad f_2: = xy^2z - xyz, \qquad f_3: = x^2y^2 - z^2.$$

The total degree ordering of power products is used in this example: first order by total degree and, within a given degree, order lexicographically. We took an example with a particularly simple structure of the polynomials in order to make the reduction process simple and to emphasize the crucial point: the difference of the crude version of the algorithm and the improved version, which is reflected in the pairs of polynomials $\{f_1, f_2\}$, for which the S-polynomials have to be considered.

A trace of the crude form of the algorithm could be as follows (if the selection strategy 1. for pairs of polynomials is used: in the trace, we write $f_i, f_j \rightarrow f_k$ for indicating that the reduction of the S-polynomial of f_i and f_j leads to the polynomial f_k):

$$f_2, f_3 \rightarrow f_4: = x^2yz - z^3,$$
$$f_1, f_4 \rightarrow f_5: = xz^3 - xz^2,$$
$$f_2, f_4 \rightarrow f_6: = yz^3 - z^3,$$
$$f_3, f_4 \rightarrow 0,$$
$$f_5, f_6 \rightarrow f_7: = xyz^2 - xz^2,$$
$$f_4, f_7 \rightarrow f_8: = z^4 - x^2z^2, \cdot$$

$$f_2, f_7 \to 0,$$
$$f_5, f_7 \to 0,$$
$$f_6, f_7 \to 0,$$
$$f_5, f_8 \to f_9: = x^3z^2 - xz^2,$$
$$f_6, f_8 \to 0.$$

The S-polynomials of all the other pairs are reduced to zero. All together one has to reduce $(9.8)/2 = 36$ S-polynomials.

In the improved algorithm, first, by ReduceAll, f_1, f_2, f_3 are reduced with respect to each other. In this example, this reduction process leaves the original basis unchanged. Then, by NewBasis, f_1, f_2, f_3 are put into G. Simultaneously the set of pairs B for which the S-polynomial have to be considered is generated. The first pair, again, is

$$f_2, f_3 \to f_4.$$

In this phase, again a call to ReduceAll is made. It is detected that, modulo $\{f_2, f_3, f_4\}$, f_1 can be reduced to f_5, hence, f_1 can be deleted from G and, correspondingly, the pairs $\{f_1, f_2\}$ and $\{f_1, f_3\}$ can be deleted from B. By NewBasis, f_4 and f_5 are adjoined to G and B is updated. The consideration of the next pair in B yields

$$f_2, f_4 \to f_6.$$

ReduceAll has no effect in this case. Thus, f_6 is adjoined to the basis immediately and B is updated. The consideration of the next pair $\{f_3, f_4\}$ in B can be skipped by application of Criterion1: $LP(f_2) = xy^2z$ divides $LCM(LP(f_3), LP(f_4)) = x^2y^2z$ and $\{f_3, f_2\}$ and $\{f_2, f_4\}$ are not in B any more, because they already were considered. The consideration of the next pairs in B yields

$$f_5, f_6 \to f_7,$$
$$f_4, f_7 \to f_8,$$

with the corresponding updating of G and B (no reductions and cancellations of polynomials in G are possible!). The S-polynomials of the next pairs reduce to zero

$$f_2, f_7 \to 0,$$
$$f_5, f_7 \to 0.$$

The criterion does not detect this fact a priori! However, the consideration of the next pair $\{f_6, f_7\}$ can, again, be skipped by application of Criterion1: f_5 is a suitable p in the criterion. Then, the following pairs are considered:

$$f_5, f_8 \to f_9,$$
$$f_6, f_8 \to 0,$$
$$f_4, f_9 \to 0.$$

The next pair $\{f_7, f_9\}$ may, again, be skipped by application of Criterion1. Finally,

$$f_5, f_9 \to 0.$$

From now on, the application of Criterion1 detects a priori, without actually carrying out the reductions, that all the remaining pairs may be skipped. Hence, instead of 36 reductions, only 11 have to be carried out with the improved algorithm. The pair $\{f_3, f_8\}$ is an example of a pair, for which Criterion2 is successful. The gain by using the criteria, in particular Criterion1, becomes more drastic as the complexity of the examples, in terms of the number of variables, the degrees of polynomials and the number of polynomials, increases.

6.5. APPLICATION: CANONICAL SIMPLIFICATION, DECISION OF IDEAL CONGRUENCE AND MEMBERSHIP, COMPUTATION IN RESIDUE CLASS RINGS

In this section, it is shown how our algorithm for constructing Gröbner bases may be applied for algorithmic solutions to the canonical simplification problem modulo polynomial ideals, the decision problems '$f \equiv_F g$' and '$f \in \text{Ideal}(F)$', and the problem of effectively computing in the associative algebra $K[x_1, \ldots, x_n]/\text{Ideal}(F)$. Actually, the three problems are intimately connected with each other. This connection is summarized in the following definitions and lemmas whose proof may be found in [6.3]. The concepts involved in these lemmas have been developed and refined in various papers by B. Caviness, J. Moses, D. Musser, H. Lausch and W. Nöbauer, R. Loos. M. Lauer, and the author: see [6.3] for a detailed reference to the literature.

Let T be an arbitrary (decidable) set (for example, $T: = K[x_1, \ldots, x_n]$) and $\sim$ an equivalence relation on T (for example, $\sim = \equiv_F$).

DEFINITION 6.7. An algorithm C with inputs and outputs in T is called a '*canonical simplifier*' (or 'ample function') for $\sim$ (on T) iff for all objects f, g in T

(SE) $C(f) \sim f$ and

(SC)   if $f \sim g$ then $C(f) = C(g)$,

(i.e. C singles out a unique representative in each equivalence class. $C(f)$ is called a *canonical form* of f).

LEMMA 6.5. $\sim$ *is decidable if there exists a canonical simplifier C for* $\sim$.
    Proof. By (SE) and (SC): $f \sim g$ iff $C(f) = C(g)$. The converse of the lemma is true, also. However, the simplification algorithm constructed in the proof of the converse is of no practical value, see [6.3], [6.4].

LEMMA 6.6. Let R be a computable (binary) operation on T, such that $\sim$ is a congruence relation with respect to R. Assume we have a canonical simplifier C for $\sim$. Define:

$$\text{Rep}(T): = \{f \epsilon T | C(f) = f\}(\text{set of '}canonical\ representatives\text{'},$$

$$\text{ample set}),$$

$$R'(f,\ g): = C(R(f,\ g))\ (\text{for all } f,\ g \epsilon \text{Rep}(T)).$$

Then, $(\text{Rep}(T), R')$ is isomorphic to $(T/\sim, R/\sim)$, $\text{Rep}(T)$ is decidable, and R' is computable. (Here, $R/\sim([f], [g]): = [R(f,\ g)]$, where $[f]$ is the congruence class of f with respect to $\sim$).   •
    Lemma 6.6 shows that, having a canonical simplifier for an equivalence relation that is a congruence with respect to a computable operation, one can *algorithmically* master the factor structure. The theorem is proven by realizing that $i(f): = [f]\ (f \epsilon \text{Rep}(T))$ defines an isomorphism between the two structures and by checking the computability properties. Applying these general concepts and facts to the case of polynomial ideals we first note:

COROLLARY 6.1 (to Theorem 6.1). *Let S be an arbitrary normal form algorithm in the sense of Definition 6.2 and F a Gröbner basis. Then $C: = \lambda f.\ S(F, f)$, i.e. the algorithm, that takes the fixed F and a variable f as input and computes $S(F, f)$, is a canonical simplifier for $\equiv_F$.*
    Proof. (SE) is fulfilled because, clearly, $f \equiv_F g$ if $f \rightarrow_F g$ (see Definition*

6.1). By iteration, $f \equiv_F S(F, f)$. (SC), in case of $\equiv_F$, is just the content of Theorem 6.1.   ●

In addition, one can prove the following lemma.

LEMMA 6.7 [6.7], [6.8]. *Let F be a Gröbner basis. Then $B := \{[u] | u$ is a power product that is not a multiple of the leading power product of any of the polynomials in $G\}$ is a linearly independent vector space basis for the vector space $K[x_1, \ldots, x_n]/Ideal(F)$ (the residue class ring modulo Ideal(F)).*

Proof. Assume that there is a linear dependence

$$c_1 \cdot [u_1] + c_2 \cdot [u_2] + \ldots + c_l \cdot [u_l] = 0$$

for some u_i in B. Then

$$f := c_1 \cdot u_1 + c_2 \cdot u_2 + \ldots + c_l \cdot u_l \epsilon Ideal(F).$$

Hence, by Theorem 6.1, f must be reducible to 0 modulo F. However, f is already in normal form because, by definition of B, non of the u_i can be reduced modulo F. Thus, $f = 0$, i.e. $c_1 = \ldots = c_l = 0$.   ●

Based on the above lemmata, the following problems can be solved by the following methods (for S use the normal form algorithm NormalForm described in Algorithm 6.1):

PROBLEM 6.3.
Given F.
Find a canonical simplifier C for the congruence $\equiv_F$ modulo Ideal(F).

METHOD 6.1 [6.12], [6.9].
Compute $G := GB(F)$.
Then the normal form algorithm $S(G, f)$ is a canonical simplifier for $\equiv_F$.

PROBLEM 6.4.
Given F, f, g.
Decide, whether $f \equiv_F g$.

METHOD 6.2 [6.9].
Compute $G := GB(F)$.
Then: $f \equiv_F g$ iff $S(G, f) = S(G, g)$.

PROBLEM 6.5.
Given E, a finite set of equations between generators of a commutative semigroup and two words f, g.
   Decide whether the equality $f = g$ is derivable from E.

METHOD 6.3 [6.19] [6.23].
Let $x_1, \ldots, x_n$ be the finitely many generators of the commutative semigroup. Conceive every equation $p = q$ in E as a polynomial $p - q$ in $Q[x_1, \ldots, x_n]$.
Compute $G: = GB(E)$.
   Then: $f = g$ is derivable from E iff $S(G, f) = S(G, g)$.

PROBLEM 6.6.
Given F, f.
   Decide whether $f \epsilon \mathrm{Ideal}(F)$.

METHOD 6.4 [6.9].
Compute $G: = GB(F)$.
   Then: $f \epsilon \mathrm{Ideal}(F)$ iff $S(G, f) = 0$.

PROBLEM 6.7.
Given F_1, F_2:
   Decide whether $\mathrm{Ideal}\ (F_1) \subseteq \mathrm{Ideal}(F_2)$.

METHOD 6.5 [6.9], [6.10].
Compute $G_2: = GB(F_2)$.
   Then: $\mathrm{Ideal}(F_1) \subseteq \mathrm{Ideal}(F_2)$ iff for all $f \epsilon F_1: S(G_2, f) = 0$.

PROBLEM 6.8.
Given F.
   Find a linearly independent basis B for the vector space $K[x_1, \ldots, x_n]/$ $\mathrm{Ideal}(F)$ (the residue class ring modulo $\mathrm{Ideal}(F)$) and,
for any two basis elements $[u]$ and $[v]$ in B find a linear representation of $[u] \cdot [v]$ in terms of the basis elements in B (i.e. find the 'multiplication table' for $K[x_1, \ldots, x_n]/\mathrm{Ideal}(F)$).

METHOD 6.6 [6.7], [6.8].
Compute G: $= GB(F)$.
Take B: $= \{[u] | u$ is a power product that is not a multiple of the leading power product of any of the polynomials in $G\}$.
$S(G, u \cdot v)$ yields a linear representation of $[u] \cdot [v]$.

PROBLEM 6.9.
Given F, f, h (where $K[x_1, \ldots, x_n]/\text{Ideal}(F)$ is assumed to be finite-dimensional as a vector space).
Find g, such that $f \cdot g \equiv_F h$ (if such a g exists).

METHOD 6.7.
Compute G: $= GB(F)$.
Represent f and h as a linear combination of the elements in B (see Method 6.6) and represent g as a linear combination with unknown coefficients. Thus, one gets a linear system of equations for the unknown coefficients, which is solvable iff a solution g exists.   •

Note, that all the above methods are 'uniform' in the sense that F is a free parameter in the respective algorithms. Thus, for example, Method 6.3 is a solution to the uniform word problem for finitely generated commutative semigroups (which is equivalent, for example, to the reachability problem for reversible Petri nets). It has been proven [6.5], [6.6] that the uniform word problem for finitely generated commutative semigroups and, also, the uniform congruence problem for polynomial ideals in $Q[x_1, \ldots, x_n]$ is exponentially space complete, i.e. is an intrinsically hard problem. Method 6.2 shows that this problem can be 'easily' reduced to the problem of constructing Gröbner bases. Hence, the problem of constructing Gröbner bases must be an intrinsically hard problem. For practice, this means that the worst case behavior of the Algorithm 6.2 and 6.3 may be extremely bad. However, this does not mean that it is useless to construct Gröbner bases, because in the particular cases at hand, the algorithm may perform well (for example, if the input F is 'nearly' a Gröbner basis). Also, if for a given F the Gröbner basis G has been constructed, an infinite number of particular algorithmic problems of the kind '$f \in \text{Ideal}(F)$?', 'compute a representation of $[u] \cdot [v]$' etc. can be solved extremely easily.

EXAMPLE 6.7. For F as in Example 6.1, $f: = xy$ is not in Ideal(F), because

$$S(GB(F), xy) = -x^2 + 1/2x \neq 0.$$

$$f \equiv_F g: = x^2y + 3/2xy + 1/2y + 3x^2 + 3/2x - 3/2,$$

because $S(GB(F), g)$ is also $-x^2 + 1/2x$.

EXAMPLE 6.8. The following reversible Petri net

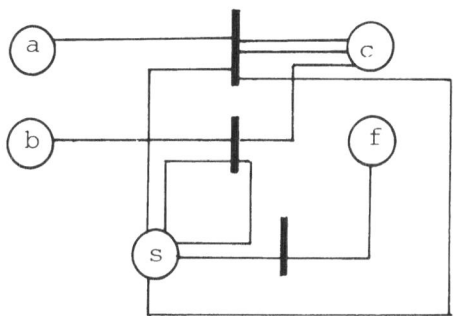

is a Petri net with places a, b, c, f, s and three transitions that may be described by the rules

$$as \rightarrow c^2s,$$

$$bs \rightarrow cs$$

$$s \rightarrow f,$$

where it is implicitly assumed that the 'reverse' rules

$$c^2s \rightarrow as$$

etc. are also available. Let

$$F = \{as - c^2s, \; bs - cs, \; s - f\}.$$

Then: a configuration v is reachable from configuration w iff $v \equiv_F w$. For example, $a^5bc^3f^2s^3$ is reachable from $a^5b^2c^2s^5$ iff $a^5bc^3f^2s^3 \equiv_F a^5b^2c^2s^5$. In order to answer such questions, we first compute (w.r.t the total degree ordering)

$$G: = GB(F) = \{s - f, \; cf - bf, \; b^2f - af\}.$$

$a^5bc^3f^2s^3$ is reachable from $a^5b^2c^2s^5$, because the normal forms of both markings are a^7f^5 (with respect to G), whereas cs^2 is not reachable from c^2s, because their respective normal forms are distinct, namely bf^2 and af.

EXAMPLE 6.9. For F of Example 6.1,

$$B = \{[1], [x], [x^2]\}$$

is a linearly independent vector basis for $K[x, y]/\text{Ideal}(F)$, see the corresponding reduced Gröbner basis G in Example 6.5.

$$[x] \cdot [x^2] = 5/2[x^2] + 5/2[x],$$

because

$$S(GB(F), x^3) = 5/2x^2 + 5/2x.$$

EXAMPLE 6.10. As an application of the construction of inverses in polynomial residue class rings, we take the simplification of radical expressions. For the formulation of the problem see [6.45]. Consider, for example, the problem of rationalizing the denominator of

$$\frac{1}{x + 2^{1/2} + 3^{2/3}}.$$

This problem may be solved by considering the given expression as an element in $Q(x)[2^{1/2}, 3^{2/3}]$, which is isomorphic to $Q(x)[y_1, y_2]/\text{Ideal}(y_1^2 - 2, y_2^3 - 3)$, i.e. the polynomial ring in the two indeterminates y_1, y_2 over the rational function field $Q(x)$ modulo the ideal generated by the polynomials $y_1^2 - 2$ and $y_2^3 - 3$. The application of the algorithm yields the equivalent Groebner-basis

$$G: = \{y_1^2 - 2, y_2^3 - 3\},$$

i.e. it is shown by the application of the algorithm that the given basis is already a Groebner-basis. (In fact, in this simple case, this can be shown by Criterion2 in Algorithm 6.3.) The residue classes of

$$1, y_1, y_2, y_1y_2, y_2^2, y_1y_2^2$$

form a vector space basis for $Q(x)[y_1, y_2]/\text{Ideal}(y_1^2 - 2, y_2^3 - 3)$. In order to obtain the iverse of $x + 2^{1/2} + 3^{2/3}$ we merely have to solve the equation

$$(x + y_1 + y_2^2) \cdot$$
$$\cdot (a_1 + a_2y_1 + a_3y_2 + a_4y_1y_2 + a_5y_2^2 + a_6y_1y_2^2) = 1.$$

By using the reductions $y_1^2 \to_G 2$, $y_2^3 \to_G 3$ this yields a *linear* system of equations in the unknowns $a_1, \ldots, a_6$ (by comparison of coefficients at the power products $1, y_1, \ldots, y_1 y_2^2$), whose solution is

$$a_1 = (x^5 - 4x^3 + 9x^2 + 4x + 18)/d,$$

$$a_2 = (-x^4 + 4x^2 + 18x - 4)/d,$$

$$a_3 = (3x^3 + 18x + 27)/d,$$

$$a_4 = (-9x^2 - 6)/d,$$

$$a_5 = (-x^4 - 9x + 4)/d,$$

$$a_6 = (2x^3 - 4x - 9)/d,$$

where $d = x^6 - 6x^4 + 18x^3 + 12x^2 + 108x + 73$.

6.6. APPLICATION: SOLVABILITY AND EXACT SOLUTION OF SYSTEMS OF ALGEBRAIC EQUATIONS

In this section, it is shown how the algorithm for constructing Gröbner bases may be used for the exact solution of systems of algebraic equations and questions about the solvability of such systems. The significance of Gröbner bases for problems in this category stems from the fact that, for Gröbner bases, the explicit construction of all the elimination ideals is extremely simple. This is particularly true for Gröbner bases with respect to the purely lexicographical ordering of power products. It is not so easy for Gröbner bases with respect to other orderings, for example, the total degree ordering. Still, it is also reasonable to construct Gröbner bases with respect to the total degree ordering for solving algebraic systems because, in extensive computational experiments, it turned out recently [6.46] that the complexity of the algorithm for constructing Gröbner bases is extremely sensitive to a permutation of variables when the purely lexicographical ordering is used, whereas it is nearly stable, when the total degree ordering is used. Furthermore, the complexity with respect to the total degree ordering is approximately in the same range as the complexity with respect to the purely lexicographical ordering, when the most favorable permutation of variables is used. Since, for a given example, there is no a priori method to predict which permutation of the variables will give the best computation times, it, therefore, is also a good method to compute the Gröbner basis with respect to the total degree ordering and then accept the disadvantage that the computation of the elimination ideals is not so easy as in the case of the purely lexicographical ordering. In the sequel, we present the method with respect to both orderings of power products.

LEMMA 6.8 [6.15]. *Let F be a Gröbner basis with respect to the purely lexicographical ordering of power products. Without loss of generality let us assume* $x_1 <_T x_2 <_T \ldots <_T x_n$. *Then*

$$Ideal(F) \cap K[x_1, \ldots, x_i] = Ideal(F \cap K[x_1, \ldots, x_i])$$

(for i = 1, . . ., n), where the ideal on the right-hand side is formed in $K[x_1, \ldots, x_i]$.   •

This lemma shows that the '*i-th elimination ideal*' of F is generated by just those polynomials in F that depend only on the variables $x_1, \ldots, x_i$.

Proof. If $f \in Ideal(F) \cap K[x_1, \ldots, x_i]$, then f can be reduced to 0 modulo F (use Theorem 6.1). With respect to the purely lexicographical ordering determined by $x_1 <_T x_2 <_T \ldots <_T x_n$, this means that f can be reduced to zero by subtraction of appropriate multiples $b_j \cdot u_j \cdot f_j$ $(f_j \epsilon F)$ such that $LP(f_j)$ contains only indeterminates from the set $\{x_1, \ldots, x_i\}$ and, hence, all power products occurring in f_j contain only indeterminates in this set. Also u_j can contain only indeterminates in this set. Adding all these $b_j \cdot u_j \cdot f_j$, one gets a representation of f of the form

$$f = \sum a_j \cdot u_j \cdot f_j$$

which shows that f is in $Ideal(F \cap K[x_1, \ldots, x_i])$.   •

PROBLEM 6.10.
Given F.
Decide, whether F has a solution (i.e. whether there exist $a_1, \ldots, a_n$ in an algebraic extension of K such that for all f in F: $f(a_1, \ldots, a_n) = 0$.)

METHOD 6.8 [6.7], [6.8].
Compute $G := GB(F)$.
Then: F *is unsolvable iff* $1 \epsilon G$.
Proof. It is well known that F has a solution iff $1 \notin Ideal(F)$, see, for example, [6.47]. Now, $Ideal(F) = Ideal(G)$ and $1 \epsilon Ideal(G)$ iff 1 is reducible w.r.t. G (by Theorem 6.1). The latter is true iff $1 \epsilon G$.

PROBLEM 6.11.
Given F.
Decide, whether F has finitely oder infinitely many solutions.

METHOD 6.9 [6.7], [6.8].
Compute $G: = GB(F)$.
Then: F has finitely many solutions iff for all i ($1 \leqslant i \leqslant n$): a power product of the form $x_i^{j_i}$ occurs among the leading power products of the polynomials in G.

Proof. It is well known that F has finitely many solutions iff the vector space $K[x_1, \ldots, x_n]$/Ideal(F) has finite vector space dimension, see, for example, [6.47]. Because of Lemma 6.7 this is true iff the set B considered in Lemma 6.7 is finite. It is easy to see from the definition of B that B is finite iff the condition stated in Method 6.9 is satisfied.

About the exact dimension of polynomial ideals, one can say more than is expressed above by using Gröbner bases for computing the Hilbert function of polynomial ideals. Many details are given in [6.33], [6.34].

PROBLEM 6.12.
Given F (solvable, with finitely many solutions).
Find all the solutions of the system F.

METHOD 6.10 [6.15].
Compute $G: = GB(F)$ with respect to the purely lexicographical ordering of power products.

The polynomials in G, then, have there variables "separated" in the precise sense of Lemma 6.8 (G is 'triangularized'). G contains exactly one polynomial of $K[x_1]$ (actually, it is the polynomial in Ideal(G) $\cap$ $K[x_1]$ with smallest degree).

The successive elimination can, then, be carried out by the following process:

$$p := \text{the polynomial in } G \cap K[x_1]$$
$$X_1 := \{(a)|p(a) = 0\}$$
$$\text{for } i := 1 \text{ to } n - 1 \text{ do}$$
$$X_{i+1} := \emptyset$$

for all $(a_1, \ldots, a_i) \epsilon X_i$ *do*

$H: = \{g(a_1, \ldots, a_i, x_{i+1})|$

$g \epsilon G \cap K[x_1, \ldots, x_{i+1}] - K[x_1, \ldots, x_i]\}$

$p: =$ greatest common divisor of the polynomials in H

(Actually, $\{p\} = GB(H)$; in the case of univariate polynomials the algorithm GB specializes to Euclid's algorithm!)

$X_{i+1}: = X_{i+1} \cup \{a_1, \ldots, a_i, a)|p(a) = 0\}$.

Upon termination, X_n will contain all the solutions. (Note that some of the p may be 1, i.e. the corresponding partial solution $(a_1, \ldots, a_i)$ can not be continued.)   •

Of course, for the univariate polynomials p occurring in the algorithm, the 'exact' determination of all their zeros may not be possible effectively. However, of course, this is not a deficiency of the particular method but an intrinsic limitation of algorithmic solvability of polynomial equations. Still, Method 6.10 is an algorithmic method (using only arithmetic in K) for completely reducing the multivariate problem to the univariate one.

Before we can give a method for Problem 6.11 that is based on Gröbner bases with respect to arbitrary orderings of power products we must solve the following problem.

PROBLEM 6.13.
Given a Gröbner basis G, such that G, as a system of equations, has only finitely many solutions.
    Find the $p \epsilon$Ideal$(G) \cap K[x_1]$ with minimal degree.

METHOD 6.11 [6.8].
(In case the purely lexicographical ordering with $x_1 <_T x_2 <_T \ldots <_T x_n$ is used, the solution of the problem is easy, see Method 6.10. In the other cases proceed by the following method.)
    Determine $d_0, \ldots, d_1$ by the following process, which involves the solution of systems of linear equations in every step:

$i: = 0$

repeat $p_i: = S(G, x_1^i)$

              $i: = i + 1$

until there exists $(d_0, \ldots, d_{i-1}) \neq (0, \ldots, 0)$ such that $d_0 \cdot p_0 + \ldots +$ $d_{i-1} \cdot p_{i-1} = 0$

$l: = i - 1$

Then, $p = d_0 \cdot 1 + d_1 \cdot x_1 + \ldots + d_l \cdot x_1^l$.

METHOD 6.12 [6.8] for solving Problem 6.12.
Compute $G: = GB(F)$.
The successive elimination can, then, be carried out by the following process:

$p: =$ the polynomial in Ideal$(G) \cap K[x_1]$ of minimal degree (see Method 6.11)

$X_1: = \{(a)|p(a) = 0\}$

for $i: = 1$ *to* $n - 1$ *do*

$X_{i+1}: = \emptyset$

for all $(a_1, \ldots, a_i) \in X_i$ *do*

$H: = \{g(a_1, \ldots, a_i, x_{i+1}, \ldots, x_n)|g \in G\}$

$H: = GB(H)$

$p: =$ the polynomial in Ideal$(H) \cap K[x_{i+1}]$ of minimal degree

$X_{i+1}: = X_{i+1} \cup \{(a_1, \ldots a_i, a)|p(a) = 0\}$

Upon termination, X_n will contain all the solutions. (Note, again, that some of the p may be one, i.e. the corresponding partial solution $(a_1, \ldots, a_i)$ can not be continued. Also, of course, one will store the Gröbner basis H corresponding to a particular partial solution $(a_1, \ldots, a_i)$ and use it instead of G for construction of H corresponding to $(a_1, \ldots, a_i, a)$.)

EXAMPLE 6.11. The system F of Example 6.1 is solvable, because $G = GB(F)$ does not contain the polynomial 1 (see Example 6.5).
    The system

$$F: = \{x^2y - x^2, x^3 - x^2 + y, xy^2 - xy + 2\}$$

is unsolvable. Let us use the total degree ordering in this example.

$$SPolynomial(x^2y - x^2, x^3 - x^2 + y) = x^2y - x^3 - y^2 \to_F$$

$$\to_F - x^3 - y^2 + x^2 \to_F - y^2 + y.$$

Thus, we have to adjoin $y^2 - y$ to the basis.

$$SPolynomial(xy^2 - xy + 2, y^2 - y) = 2,$$

which can not be reduced further. Hence, we have to adjoin 1 to the basis. This is the signal that F is unsolvable.

EXAMPLE 6.12. F of Example 6.1 has only finitely many solutions, because x^3 and y appear as leading power products in $GB(F)$.

$$F: = \{x^2y - y^2 - x^2 + y, x^2 - y\}$$

has infinitely many solutions. Actually, F is already a Gröbner basis (with respect to the total degree ordering of power products): check by applying Algorithm 6.3 which, in this case, does not adjoin any new polynomial to F. No power products of the form y^j occurs among the leading power products. Hence, F has infinitely many solutions.

EXAMPLE 6.13. For F of Example 6.1,

$$GB(F) = \{x^3 - 5/2x^2 - 5/2x, y + x^2 - 3/2x - 3\}.$$

The solutions a of the first (univariate!) polynomial are 0, $(5 + \sqrt{65})/4$, $(5 - \sqrt{65})/4$. Each of these solutions can be continued to a solution (a, b) of F by solving the second polynomial in the form $y + a^2 - 3/2a - 3$ for y. This yields $(0, 3)$, $((5 + \sqrt{65})/4, -(3 + \sqrt{65})/4)$, $((5 - \sqrt{65})/4, (-3 + \sqrt{65})/4)$ as the three solutions of the system.

EXAMPLE 6.14. The same example can also be treated by Method 6.12. With respect to the total degree ordering, $G: = GB(F) = \{g_1, g_2, g_3\}$ where

$$g_1: = x^2 + y - 3/2x - 3,$$
$$g_2: = xy - y + x + 3,$$
$$g_3: = y^2 - 5/2y - 4x - 3/2.$$

We now compute the normal forms of $1, x, x^2, \ldots$:

$$S(G, 1) = 1,$$
$$d_0 \cdot 1 = 0 \text{ has no non-trivial solution.}$$
$$S(G, x) = x,$$
$$d_0 \cdot 1 + d_1 \cdot x = 0 \text{ has no non-trivial solution.}$$

$S(G, x^2) = -y + 3/2x + 3,$

$d_0 \cdot 1 + d_1 \cdot x + d_2 \cdot x^2 = 0$ has no non-trivial solution.

$S(G, x^3) = -5/2y + 25/4x + 15/2,$

$d_0 \cdot 1 + d_1 \cdot x + d_2 \cdot x^2 + d_3 \cdot x^3 = 0$ leads to the following linear system of equations:

$$
\begin{aligned}
-5/2d_3 - d_2 &= 0, \\
25/4d_3 + 3/2d_2 + d_1 &= 0, \\
15/2d_3 + 3d_2 + d_0 &= 0,
\end{aligned}
$$

which has (after normalization $d_3 = 1$) the unique solution $d_3 = 1$, $d_2 = -5/2, d_1 = -5/2, d_0 = 0$. This means that

$$p: = x^3 - 5/2x^2 - 5/2x$$

is the polynomial in Ideal(G) $\cap$ $K[x]$ with minimal degree (in accordance to what we already have seen in Example 6.13). p has the three solutions $a_1 = 0, a_2 = (5 + \sqrt{65})/4, a_3 = (5 - \sqrt{65})/4$. Substitution of a_1 yields

$$
\begin{aligned}
g_1(a_1) &= y - 3, \\
g_2(a_1) &= -y + 3, \\
g_3(a_1) &= y^2 - 5/2y - 3/2.
\end{aligned}
$$

The Gröbner basis corresponding to these three polynomials is

$$G': = \{y - 3\}.$$

By computing the normal forms $1, y, y^2, \ldots$ and looking at the corresponding systems of linear equations as above one detects that

$$p': = y - 3$$

is the polynomial in Ideal(G') $\cap$ $K[y]$ of minimal degree. Of course, in this particularly simple example, this can be seen immediately from the Gröbner basis. Hence, (a_1, b_1) with $b_1: = 3$ is the first solution of the system. Similarly, substitution of a_2 yields

$$
\begin{aligned}
g_1(a_2) &= y + (3 + \sqrt{65})/4, \\
g_2(a_2) &= (1 + \sqrt{65})/4y + (17 + \sqrt{65})/4, \\
g_3(a_2) &= y^2 - 5/2y - (13 + \sqrt{65})/2.
\end{aligned}
$$

The Gröbner basis corresponding to these three polynomials is

$$G'': = \{y + (3 + \sqrt{65})/4\} \quad \text{and} \quad p'': = y + (3 + \sqrt{65})/4$$

is the polynomial in $\text{Ideal}(G'') \cap K[y]$ of minimal degree. Hence, (a_2, b_2) with $b_2: = -(3 + \sqrt{65})/4$ is the second solution of the system.

Finally, substitution of a_3 yields, again, three polynomials in $K[y]$ whose Gröbner basis consists of the polynomial $y + (3 - \sqrt{65})/4$. Hence, the third solution is (a_3, b_3) with $b_3: = (-3 + \sqrt{65})/4$.

EXAMPLE 6.15. Given F consisting of

$$4x^2 + xy^2 - z + 1/4,$$

$$2x + y^2z + 1/2,$$

$$x^2z - 1/2x - y^2,$$

the corresponding Gröbner basis G (with respect to the purely lexico-graphical ordering, where $z <_T y <_T x$) consists of

$$z^7 - 1/2z^6 + 1/16z^5 + 13/4z^4 + 75/16z^3 + 171/8z^2 +$$
$$+ 133/8z - 15/4,$$

$$y^2 - 19188/497z^6 + 318/497z^5 - 4197/1988z^4 -$$
$$- 251555/1988z^3 - 481837/1988z^2 +$$
$$+ 1407741/1988z - 297833/994,$$

$$x + 4638/497z^6 - 75/497z^5 + 2111/3976z^4 +$$
$$+ 61031/1988z^3 + 232833/3976z^2 - 85042/497z +$$
$$+ 144407/1988.$$

Applying Method 6.10 for solving G, one first had to find all the solutions of the first polynomial, which is univariate. Each of these solution a_1, can be continued to two solutions (a_1, a_2) of the second polynomial and each of these (a_1, a_2) can be continued to a solution (a_1, a_2, a_3) of the third polynomial. The solutions of the first polynomial can be determined systematically with any guaranteed precision, see [6.48]. It has not yet been studied systematically how, numerically, the precision of the solutions of the first equation must be fixed in order to guarantee a given precision for all the solutions of the last equation. This is a near-at-hand important problem for future study.

EXAMPLE 6.16. Sometimes, it is necessary to solve systems of algebraic equations with 'symbolic' coefficients. For example consider F consisting of

$$f_1: = x_4 + (b - d),$$
$$f_2: = x_4 + x_3 + x_2 + x_1 + (-a - c - d),$$
$$f_3: = x_3x_4 + x_1x_4 + x_2x_3 + (-ad - ac - cd),$$
$$f_4: = x_1x_3x_4 + (-acd),$$

where $x_1 <_T x_2 <_T x_3 <_T x_4$ are the polynomial indeterminates and a, b, c, d are 'symbolic' coefficients. One might like to solve this system for x_1, x_2, x_3, x_4. This is nothing else then saying that one conceives the polynomials as elements in $Q(a, b, c, d)[x_1, \ldots, x_4]$, where $Q(a, b, c, d)$ is the field of rational functions over Q. Our algorithm works over arbitrary fields and, hence, in particular also over $Q(a, b, c, d)$. Some steps of Algorithm 6.3 are:

Reduction of f_1 modulo f_2 (by subtraction of f_2 from f_1 and normalizing the coefficient of the leading power product to 1) yields

$$f_1': = x_3 + x_2 + x_1 + (- a - b - c) \ (f_1 \text{ may be canceled}).$$

Reduction of f_2 modulo f_1' yields

$$f_2': = x_4 + (b - d) \ (f_2 \text{ may be canceled}).$$

Reduction of f_3 modulo the other polynomials (starting with the subtraction of $x_3 \cdot f_2'$ and, then executing several other reduction steps) yields

$$f_3': = x_2^2 + 2x_1x_2 - (a + 2b + c - d) x_2 + x_1^2 -$$
$$- (a + b + c) x_1 + (ab + ac + b^2 + bc - bd)$$
$$(f_3 \text{ may be canceled}).$$

Reduction of f_4 yields

$$f_4': = x_1x_2 + x_1^2 - (a + b + c) x_1 - acd/(b - d) \ (\text{cancel } f_4).$$

(Note here that division in $Q(a, b, c, d)$ has to be performed. f_3' can now be further reduced (using f_4') yielding f_3''

$$f_3'': = x_2^2 - (a + 2b + c - d) x_2 - x_1^2 + (a + b + c) x_1 +$$
$$+ (ab^2 + abc - abd + acd + b^3 + b^2c -$$
$$- 2b^2d - bcd + bd^2)/(b - d).$$

Cancel f'_3. No further reduction is possible. Therefore, we consider

$$SPolynomial(f''_3, f'_4) = x_1 \cdot f''_3 - x_2 \cdot f'_4.$$

Reduction of this polynomial yields

$$f_5: = x_2 + (b^2 - 2bd + d^2)/(acd)\ x_1^2 +$$
$$+ (abc + abd - ad^2 + bcd - cd^2)/(acd)\ x_1 + (-b + d).$$

Now, again, a number of reductions are possible yielding, finally,

$$g_1: = x_3 + (-b^2 + 2bd - d^2)/(acd)\ x_1^2 +$$
$$+ (-abc + abd + acd + ad^2 - bcd + cd^2)/(acd)\ x_1 +$$
$$+ (-a - c - d),$$

$$g_2: = x_4 + (b - d),$$

$$g_3: = x_1^3 + (ac + ad + cd)/(b - d)\ x_1^2 +$$
$$+ (a^2cd + ac^2d + acd^2)/(b^2 - 2bd + d^2)\ x_1 +$$
$$+ (a^2c^2d^2)/(b^3 - 3b^2d + 3bd^2 + d^3),$$

$$g_4: = x_2 + (b^2 - 2bd + d^2)/(acd)\ x_1^2 +$$
$$+ (abc + abd - ad^2 + bcd - cd^2)/(acd)\ x_1 + (-b + d).$$

By Criterion1, the reduction of the S-polynomials of these polynomial may be skipped. Hence, $G: = \{g_1, \ldots, g_4\}$ is the reduced Gröbner basis. By Methods 6.8 and 6.9 it can be seen that the system has finitely many solutions. The system must contain a univariate polynomial in $Q(a, b, c, d)[x_1]$: g_3. A particular solution of g_3 is

$$a_1: = (-ad)/(b - d),$$

which can be extended to a solution (a_1, a_2, a_3, a_4) of the entire system, where

$$a_2: = (ab + b^2 - bd)/(b - d),$$
$$a_3: = c,$$
$$a_4: = -b + d.$$

Dividing g_3 by $(x_1 - a_1)$ one gets a quadratic polynomial whose solutions can be extended to solutions of the entire system in the same way as before.

6.7. APPLICATION: SOLUTION OF LINEAR HOMOGENEOUS EQUATIONS WITH POLYNOMIAL COEFFICIENTS

In this section, it is shown how the algorithm for constructing Gröbner bases may be used for determining a finite set of generators for all the polynomial solutions of a linear homogeneous equation with polynomial coefficients. Before the method can be described, it must be shown how one can find a linear representation of the polynomials in a basis F in terms of the polynomials in its corresponding Gröbner basis G and vice versa.

PROBLEM 6.14.
Given a Gröbner basis $G = \{g_1 \ldots, g_m\}$ and some f.
Find $h_1, \ldots, h_m$ such that $f = h_1 \cdot g_1 + \ldots + h_m \cdot g_m$ (and $LP(h_i \cdot g_i)$ $\leqslant_T LP(f)$ for $i = 1, \ldots, m$).

METHOD 6.13.
Roughly, reduce f to zero modulo G and collect the multiples of the g_i necessary in the reduction. In more detail: take Algorithm 6.1 (the normal form algorithm) and insert instructions that collect the multiples of the g_i used in the reduction.

$h_1: = \ldots := h_m: = 0$

while $f \neq 0$ do

        choose i, b, u such that $f \rightarrow_{g_i, b, u}$ and $u \cdot LP(g_i)$
        is maximal w.r.t.

        $f : = f - b \cdot u \cdot g_i$
        $h_i: = h_i + b \cdot u$

PROBLEM 6.15.
Given $F = \{f_1, \ldots, f_l\}$ and $G = \{g_1, \ldots, g_m\}$ such that $G = GB(F)$.
Find Y such that Y is a matrix of polynomials with m rows and 1 columns and

$$f_j = \sum_{1 \leqslant i \leqslant m} g_i \cdot Y_{i,j} \quad (\text{for } j = 1, \ldots, l).$$

METHOD 6.14.
The j-th column of Y consists of $h_1, \ldots, h_m$ that are obtained by the Method 6.13 for the representation of f_j ($j = 1, \ldots, l$).

PROBLEM 6.16.
Given: $F = \{f_1, \ldots, f_l\}$.
Find $G = \{g_1, \ldots, g_m\}$ and X such that $G = GB(F)$, X is a matrix of polynomials with l rows and m columns and

$$g_i = \sum_{1 \leq j \leq l} f_j \cdot X_{j, i} \quad \text{(for } i = 1, \ldots, m).$$

METHOD 6.15.
Augment Algorithm 6.2 or Algorithm 6.3 by instructions that keep track of the multiples of f_j that are used in the reduction of those polynomials whose normal form is adjoined to the basis G (compare Method 6.13)

PROBLEM 6.17.
Given a reduced Gröbner basis $G = \{g_1, \ldots, g_m\}$.
Find a matrix R with m columns such that the finitely many rows of R constitute a set of generators for the linear homogeneous equation

$$h_1 \cdot g_1 + \ldots + h_m \cdot g_m = 0 \; (h_1, \ldots, h_m \in K[x_1, \ldots, x_n]),$$

i.e. R should consist of m-tupels $(k_{1, 1}, \ldots, k_{1, m}), \ldots, (k_{r, 1}, \ldots, k_{r, m})$ of polynomials such that

$$k_{j, 1} \cdot g_1 + \ldots + k_{j, m} \cdot g_m = 0 \; (\text{for } j = 1, \ldots r)$$
and for all $(h_1, \ldots, h_m)$ for which
$$h_1 \cdot g_1 + \ldots + h_m \cdot g_m = 0$$

there exist polynomials $p_1, \ldots, p_r$ such that

$$(h_1, \ldots, h_m) =$$
$$= p_1 \cdot (k_{1, 1}, \ldots k_{1, m}) + \ldots + p_r \cdot (k_{r, 1}, \ldots, k_{r, m}).$$

METHOD 6.16 [6.14], [6.18], [6.21], [6.28], [6.33], [6.34].
$R: =$ empty matrix

for all pairs (i, j) $(1 \leq i < j \leq 1)$:

Consider $h: = SPolynomial(g_i, g_j) = u_i \cdot g_i - (c_i/c_j) \cdot u_j \cdot g_j$, where c_i is the leading coefficients of g_i, u_i is such that $s_i \cdot u_i$ is the $LCM(s_1, s_2)$, s_i is the leading power product of g_i $(i = 1, 2)$.
Reduce h to zero modulo G and store the multiples of the $g_1, \ldots, g_l$

necessary for this reduction. This gives a representation of h of the form

$$h = k_1 \cdot g_1 + \ldots + k_l \cdot g_l \text{ (compare Method 6.13!)}.$$

Add $(\ldots, u_i, \ldots, -(c_i/c_j) \cdot u_j, \ldots) - (k_1, \ldots, k_l)$ as last row in R

       $\uparrow$ $\uparrow$

   position i position j

PROBLEM 6.18.
Given $F = \{f_1, \ldots, f_l\}$ arbitrary.
Find a matrix Q with l columns such that the finitely many rows of Q constitute a set of generators for the linear homogeneous equation

$$h_1 \cdot f_1 + \ldots + h_l \cdot f_l = 0 \ (h_1, \ldots, h_l \in K[x_1, \ldots, x_n]).$$

METHOD 6.17 [6.18].
By Method 6.15, compute $G = GB(F) = \{g_1, \ldots, g_m\}$ and a matrix X with l rows and m columns such that

$$g_i = \sum_{1 \leq j \leq l} f_j \cdot X_{j,i} \quad \text{(for } i = 1, \ldots, m).$$

By Method 6.14, compute a matrix Y with m rows and l columns such that

$$f_j = \sum_{1 \leq i \leq m} g_i \cdot Y_{i,j} \quad \text{(for } j = 1, \ldots, l).$$

By Method 6.16 compute a matrix R with m columns such that the r rows of R constitute a set of generators for the linear homogeneous equations

$$h_1 \cdot g_1 + \ldots + h_m \cdot g_m = 0.$$

Then,

$$Q = \begin{pmatrix} I - Y^t \cdot X^t \\ \cdots\cdots\cdots\cdots \\ R \cdot X^t \end{pmatrix} \text{ (a block matrix)}$$

(I is the unit matrix with l rows and columns, X^t is the transposed of X).

EXAMPLE 6.17.
Let $F: = \{f_1, f_2, f_3\}$, where

$$f_1: = x^2y - xy, \qquad f_2: = xy^2 - x^2,$$
$$f_3: = x^3y - x^2y + x^3 - x^2.$$

We use the total degree ordering. First, $G: = GB(F)$ has to be computed with simultaneous determination of the matrix X. We start with a reduction of f_3:

$$f_3 - x \cdot f_1 = x^3 - x^2 = :f_3'.$$

The representation

$$f_3' = (-x) \cdot f_1 + 0 \cdot f_2 + 1 \cdot f_3$$

must be stored. Then we reduce the S-polynomial of f_1 and f_2:

$$h: = SPolynomial(f_1, f_2) = y \cdot f_1 - x \cdot f_2,$$
$$h + f_2 - f_3' = 0 = :f_4.$$

If f_4 was not zero, the following representation of f_4 in terms of f_1, f_2 and f_3 could be obtained from this reduction:

$$f_4 = y \cdot f_1 - x \cdot f_2 + f_2 - f_3 + x \cdot f_1 =$$
$$= (y + x) \cdot f_1 + (- x + 1) \cdot f_2 + (-1) \cdot f_3.$$

This example of a reduction should suffice to demonstrate how the linear representations of the new polynomials in G in terms of the polynomials in F can be obtained in general. Since, however, f_4 is zero, nothing has to be adjoined to G in this stage of the algorithm. The S-polynomial of f_1 and f_3' and also the S-polynomial of f_2 and f_3' reduce to zero. Hence,

$$G: = \{g_1, g_2, g_3\},$$

where

$$g_1: = f_1, g_2: = f_2, g_3: = x^3 - x^2,$$

is the reduced Gröbner basis corresponding to F and

$$X: = \begin{pmatrix} 1 & 0 & -x \\ 0 & 1 & 0 \\ 0 & 0 & 1 \end{pmatrix}$$

is the transformation matrix.

The matrix Y for the reverse transformation (i.e. the linear representation of the elements of F in terms of the elements in G) is obtained by Method 6.14:

f_1 reduces to zero modulo G by subtraction of g_1,
f_2 reduces to zero modulo G by subtraction of g_2,
f_3 reduces to zero modulo G by subtraction of $x \cdot g_1$ and g_3.

Hence,

$$Y: = \begin{pmatrix} 1 & 0 & x \\ 0 & 1 & 0 \\ 0 & 0 & 1 \end{pmatrix}.$$

For getting R, we have to reduce the S-polynomials of the pairs (g_i, g_j):

$$h_{1,2}: = \text{SPolynomial}(g_1, g_2) = y \cdot g_1 - x \cdot g_2.$$
$$h_{1,2} + g_2 - g_3 = 0.$$
$$h_{1,3}: = \text{SPolynomial}(g_1, g_3) = x \cdot g_1 - y \cdot g_3.$$
$$h_{1,3} = 0.$$
$$h_{2,3}: = \text{SPolynomial}(g_2, g_3) = x^2 \cdot g_2 - y^2 \cdot g_3.$$
$$h_{2,3} - y \cdot g_1 + x \cdot g_3 - g_2 + g_3 = 0.$$

From the first reduction:

$$y \cdot g_1 - x \cdot g_2 + g_2 - g_3 =$$
$$(y) \cdot g_1 + (-x + 1) \cdot g_2 + (-1) \cdot g_3 = 0.$$

Hence, the first row in R is the solution (the 'syzygy')

$$(y, -x + 1, -1).$$

The other rows of R are obtained analogously:

$$R = \begin{pmatrix} (y) & (-x + 1) & (-1) \\ (x) & (0) & (-y) \\ (-y) & (x^2 - 1) & (-y^2 + x + 1) \end{pmatrix}.$$

Finally, the computation of Q requires only some matrix multiplications: First, we note that $Y' \cdot X' = I$ in this particular example. Hence,

$$Q = \begin{pmatrix} I - Y' \cdot X' \\ \dots\dots\dots\dots \\ R \cdot X' \end{pmatrix} =$$

$$\begin{pmatrix} (0) & (0) & (0) \\ (0) & (0) & (0) \\ (0) & (0) & (0) \\ (y + x) & (-x + 1) & (-1) \\ (x + xy) & (0) & (-y) \\ (xy^2 - x^2 - y - x) & (x^2 - 1) & (-y^2 + x + 1) \end{pmatrix}$$

Of course, the first three rows can be canceled in this particular example, the last three rows constitute a complete set of generators for the solutions (h_1, h_2, h_3) to the equation $h_1 \cdot f_1 + h_2 \cdot f_2 + h_3 \cdot f_3 = 0$. •

For $K[x_1, \ldots, x_n]$-modules, as for example the module of all the solutions to the above linear equation, a notion of 'Gröbner bases' and 'reduced Gröbner bases' can be introduced, see [6.28], [6.33], [6.34]. Then the matrices Q can be reduced to a minimal set of generators and the construction can be carried over to obtain the whole 'chain of syzygies' or the 'free resolution' of a polynomial ideal.

6.8. GRÖBNER BASES FOR POLYNOMIAL IDEALS OVER THE INTEGERS

The concept of Gröbner bases, the essential properties of Gröbner bases and the algorithm for constructing Gröbner bases as reflected by Definitions 6.2, 6.3, 6.5, 6.6, Lemmata 6.1, 6.2, 6.3, Theorems 6.1, 6.2, 6.3, 6.4, Algorithms 6.1, 6.2, 6.3 and most of the applications in Sections 6.5 and 6.7 can be carried over to polynomial ideals in $Z[x_1, \ldots, x_n]$ and, in fact, to ideals in certain other rings, see [6.30]. However, a subtle analysis of the notion of reduction and, more essentially, of the notion of 'S-polynomial' must be carried out for this purpose. We can not go into the details of the theoretical foundations of the algorithm for integer polynomials. Rather, we explain the steps of the generalized algorithm in the style of the preceding sections.

The problem of deciding ideal membership for ideals in $Z[x_1, \ldots, x_n]$, the simplification problem for these ideals and related problems have a long and interesting history. For some of the details of the history, see [6.49]. The first general solution of both the simplification and (hence,) the membership problem, was given by Lauer [6.11] based on the Gröbner bases approach but needing two different types of 'S-polynomials'. Other solutions based on the Gröbner bases approach, but destroying the simple structure of the algorithm, were given in [6.15], [6.18], [6.21]. The first general solution based on a different approach was given only in [6.49]. Our own solution [6.30], which will be presented here, seems to be much more concise than the solutions given so far and leaves the simple structure of the algorithm untouched.

In addition to some ordering of the power products, in the case of $Z[x_1, \ldots, x_n]$, one also must fix some ordering of the integers, for example, $0 < -1 < 1 < -2 < 2 < -3 < 3 < \ldots$ (An axiomatic characterization of the admissible orderings is possible but will not be used in this paper). The crucial difference, then, to the case of polynomials with field coefficients is that, in the definition of 'reduction' (Definition 6.1) it is not possible to totally cancel Coefficient(g, t), where $t = u \cdot$ LeadingPowerProduct(f), because the element Coefficient(g, t)/ LeadingCoefficient(f), in general, will not be in Z. In the following, the typed variables a, b, c, d will be used for integers instead of field elements, f, g, h, k, p, q will be used for polynomials in $Z[x_1, \ldots, x_n]$, and F, G for finite sets in $Z[x_1, \ldots, x_n]$.

DEFINITION 6.8 [6.30].
$g \rightarrow_F h$ (read: 'g reduces to h modulo F') iff there exists $f\epsilon F$, b and u such that

$$g \rightarrow_{f, b, u} \quad \text{and} \quad h = g - b \cdot u \cdot f \cdot$$

$g \rightarrow_{f, b, u}$ (read: 'g is reducible using f, b, u') iff

$$a \neq 0 \quad \text{and} \quad a - b \cdot c < a,$$

where,

$$a = \text{Coefficient}(g, u \cdot \text{LeadingPowerProduct}(f)), \text{ and}$$
$$c = \text{LeadingCoefficient}(f)$$

EXAMPLE 6.18. The b in Definition 6.8 can be determined by the following algorithm $M(a, c)$, for example:

$$M(a, c): = \text{ if } a \text{ and } c \text{ have the same sign}$$
$$\text{then if } a - c < a \text{ then } M(a - c, c) + 1$$
$$\text{else } 0$$
$$\text{else if } a + c < a \text{ then } M(a + c, c) - 1$$
$$\text{else } 0$$

In practice, M may be realized by a modified integer division.   •
   The definitions, theorems, algorithms and lemmata of Section 2 can now be carried over without any change: In particular, we have again the algorithm NormalForm that produces a normal form for every polynomial, we have the notion of a Gröbner basis, the characterizations (GB2) and (GB3) of Gröbner bases and the connection between reduction and ideal congruence stated in Lemma 6.3. For the formulation of

the algorithm that constructs Gröbner bases, however, we need some additional preparation.

DEFINITION 6.9 [6.30].
The *least common reducible* of c_1, c_2 is defined as follows:

$$LCR(c_1, c_2): = \max(L(c_1), L(c_2)) \text{ (max taken w.r.t. } <),$$

where

$$L(c) \qquad : = abs(c)/2, \qquad\qquad \text{if } c \text{ is even}$$
$$- (abs(c) + 1)/2, \text{ if } c \text{ is odd.}$$

DEFINITION 6.10 [6.30].
p_1 and p_2 constitute the *critical pair* corresponding to f_1 and f_2 iff

$$p_i = a \cdot U - M(a, c_i) \cdot u_i \cdot f_i, \text{ where}$$
$$U = LCM(s_1, s_2),$$
$$a = LCR(c_1, c_2).$$
$$s_i = \text{LeadingPowerProduct}(f_i),$$
$$c_i = \text{LeadingCoefficient}(f_i),$$
$$u_i \text{ is such that } u_i \cdot s_i = U \qquad (i = 1, 2). \qquad\bullet$$

The difference of the two components of a critical pair is the analogue to the *S*-polynomial in the case of field coefficients. We formulate the algorithm for critical pairs instead of *S*-polynomials, because, at present, we do not have a formal proof that, in fact, the algorithm below is correct with *S*-polynomials instead of critical pairs, although it is very likely. Also, we would like to introduce the concept of a critical pair to the reader, because this concept may be applied to domains without any operation of subtraction also. See [6.3] for an introduction to 'critical-pair/completion' algorithms.

EXAMPLE 6.19.
$0, -1, 1, -2, 2, -3, 3, -4, 4$ are the values of L for the arguments $0, 1, 2, 3, 4, 5, 6, 7, 8$, respectively, and $LCR(3, 1) = -2$, $LCR(7, 8) = 4$. Note that $L(c) = L(-c)$. $\bullet$

The main theorem of Section 3, which gives an algorithmic characterization of Gröbner bases, and the main algorithm for the main problem can now be carried over in the following form:

THEOREM 6.6 (Buchberger [6.30]).
Let S be an arbitrary normal form algorithm. The following properties are equivalent:

(GB1) F is a Gröbner basis.

(GB3)   For all $f_1, f_2 \epsilon F, p_1, p_2$:

if p_1 and p_2 constitute the critical pair corresponding to f_1, f_2, then $S(F, f_1) = S(F, f_2)$.

PROBLEM 6.19.
Given F.
Find G, such that Ideal(F) = Ideal(G) and G is a Gröbner basis.

ALGORITHM 6.4 (Buchberger [6.30]) for solving Problem 6.19.

$G: = F$

$B: = \{f_1, f_2\}|f_1, f_2 \epsilon G\}$

while $B \neq \emptyset$ do

$\{f_1, f_2\} \quad : = a$ pair in B

$(p_1, p_2): = $ the critical pair corresponding to f_1, f_2

$(p_1', p_2'): = (S(G, p_1), S(G, p_2))$

$h' \qquad : = p_1' - p_2'$

if $h' \neq 0$ then

$B: = B \cup \{\{g, h'\}|g \epsilon G\}$

$G: = G \cup \{h'\}$.   •

Also the various improvements of the algorithm, the notion of reduced Gröbner bases and the theorem on the uniqueness of the reduced Gröbner bases (Section 3) can be carried over. We do not explicitly state the details.

EXAMPLE 6.20. Take F as in Example 6.1. Note that the leading coefficients of the polynomials in F can not be simply set to 1 by dividing the whole polynomial: the ideal would change! We fix the 'purely lexicographical' ordering for the bivariate power products with the ordering $x <_T y$ of the two indeterminates. In order to 'complete' F by Algorithm 6.4, one has to consider the 'critical pairs' of polynomials in F. We start with f_2, f_3: $LC(f_2) = 2$, $LC(f_3) = 1$, $LCR(2, 1) = 1$, $LCM(LP(f_2), LP(f_3)) = x^3y$. Thus, x^3y is the monomial that has to be reduced in one step modulo f_2 and f_3 in order to get the critical pair corresponding to f_2, f_3. The polynomial x^3y may be reduced by f_2 in the following way:

$$x^3y \to_{f_2} - x^3y + xy + y - 6x^3 + 2x^2 + 3x - 3 = :p.$$

p may be further reduced modulo f_3:

$$p \to_{f_3} x^2y + xy + y - 3x^3 + 4x^2 + 3x - 3 = :p'.$$

p' is irreducible with respect to F. The polynomial x^3y may also be reduced by f_3:

$$x^3y \to_{f_3} - x^2y - 3x^3 - 2x^2 = :q.$$

Also q is irreducible with respect to F. $p' \neq q$ and, hence,

$$f_4: = p' - q = 2x^2y + xy + y + 6x^2 + 3x - 3$$

must be adjoined to the basis.

Similarly, one now has to consider the next critical pair, for example, the one corresponding to f_1, f_4: $- 2x^2y$ is the 'least common reducible' of f_1 and f_4, which has to be reduced in one step modulo f_1 and f_4, yielding

$$p: = x^2y + 2xy + y + 9x^2 + 5x - 3 \quad \text{and}$$

$$q: = xy + y + 6x^2 + 3x - 3,$$

respectively. Reduction to normal forms yields

$$p': = -x^2y + xy + 3x^2 + 2x \text{ (using } f_4 \text{) and}$$

$$q': = xy + y + 6x^2 + 3x - 3.$$

Thus, the difference of these two polynomials must be adjoined to the basis:

$$f_5: = -x^2y - y - 3x^2 - x + 3.$$

Similarly, the consideration of the critical pair of f_4 and f_5 leads to

$$f_6: = -xy + y - x - 3.$$

The consideration of the critical pair of f_5 and f_6 leads to

$$f_7: = 2y + 2x^2 - 3x - 6.$$

Finally, the consideration of the critical pair of f_6 and f_7 leads to

$$f_8: = 2x^3 - 5x^2 - 5x.$$

The consideration of all the other critical pairs leads to identical normal forms. Hence, $G: = \{f_1, \ldots, f_8\}$ is a Gröbner basis corresponding to F. Actually, the consideration of most of these critical pairs can be avoided a prior by the improved version of the algorithm. Furthermore, some of the polynomials in the basis can also be canceled in the course of the algorithm. Reduction of all the f_i modulo $G - \{f_i\}$ leaves us with the reduced Gröbner basis $G': = \{f'_6, f_7, f_8\}$. where

$$f'_6: = -xy - y - 2x^2 + 2x + 3.$$

Note that the reduced Gröbner bases corresponding to F are different depending on whether we work in $Q[x_1, \ldots, x_n]$ or in $Z[x_1, \ldots, x_n]$.

6.9. OTHER APPLICATIONS

A number of other applications of Gröbner bases have been reported in the literature: decision, whether a given polynomial ideal is principal [6.8], Hilbert functions of polynomial ideals [6.7], [6.28], [6.33], [6.34], Lasker-Noether decomposition of polynomial ideals [6.13], free resolutions of polynomial ideals and syzygies (a generalization of the above linear equation problem with polynomial coefficients) [6.28], [6. 34], multidimensional integration [6.50] and bijective enumeration of polynomial ideals. The latter problem asks for an algorithm that enumerates bases for ideals in $R[x_1, \ldots, x_n]$ (R a ring) such that every ideal is represented exactly once in the enumeration. By Theorem 6.4, it is clear that a bijective enumeration of all ideals in $K[x_1, \ldots, x_n]$ and $Z[x_1, \ldots, x_n]$ can be achieved by bijectively enumerating all Gröbner bases in these polynomial rings, which is easily possible (see [6.37]). The applicability of Gröbner bases to other problems is investigated, for example, to the construction of Hensel codes for rational functions [6.51].

6.10. SPECIALIZATIONS, GENERALIZATIONS, IMPLEMENTATIONS, COMPLEXITY

The algorithm for constructing Gröbner bases *specializes* to Gauß' algorithm in case F consists only of linear polynomials, it specializes to Euclid's algorithm in case F consists only of univariate polynomials, it specializes to an algorithm for the word problem for finitely generated commutative semigroups in case F consists only of polynomials of the form $u - v$ (differences of power products) [6.19], [6.23]. The algorithm for $Z[x_1, \ldots, x_n]$ specializes to Euclid's algorithm in Z in case $n = 0$, [6.30].

The algorithm has been *generalized* for polynomials over various rings, in particular, over Z [6.11], [6.15], [6.18], [6.21], [6.30], and for associative algebras [6.17]. The Knuth-Bendix generalization [6.38] was already discussed in the introduction. Recently, an interesting generalization was also undertaken by G. Bauer [6.24], who gives an axiomatic definition of the concept of 'substitution' and is able to define the notion of 'critical pair' in this general context.

The algorithm has been *implemented* various times, [6.7], [6.13], [6.16], [6.21]. [6.16] is an implementation in SAC–1. R. Gebauer and H. Kredel [6.46], Univ. of Heidelberg, F.R.G., work on the implementation of the algorithm in SAC–2, which will be included in the next release of SAC–2 (announced for December 1983). SAC–2 is a large software system for symbolic computation in algebraic domains, in particular in polynomial domains. It is written in the ALDES language, whose compiler is written in FORTRAN. Thus, SAC–2 is installed easily whenever FORTRAN is available. G. E. Collins (University of Wisconsin-Madison, Departments of Computer Science) and R. Loos (Universität Karlsruhe, Institut für Informatik I) are the authors of the SAC–2 system. The implementation of our algorithm in SAC–2 by R. Gebauer and H. Kredel gives the user the choice to use various orderings of power products, to work over various coefficient domains (including the field of rational functions over Q) and to communicate in convenient input and output format with the computer.

Various analyses of the *complexity* of the algorithm have been carried out: [6.7], [6.19], [6.29], [6.6], [6.31]. Summarizing, these analyses show that the degrees of the polynomials in the reduced Gröbner bases, with probability 1, stay below $d_1 + \ldots + d_l - n + 1$, where the d_i are the degrees of the input polynomials. In exceptional cases, this bound does not hold. Many theoretical questions remain open. Typical running times in SAC–2 on an IBM 370/168: several seconds for F with 3 polynomials of degree 3 in 3 variables, 20 sec for the example in [6.15] with 6 polynomials of degree 3 in 6 variables. However, this computing time may drastically change if a different permutation of the variables and purely lexicographical ordering is used. For the worst permutation, the computation was as high as 10 000 sec, whereas in the total degree ordering the computation time for the same example was always in the range 20–30 sec independent of the permutation of variables. See Section 6 for the consequences of these obsevations.

ACKNOWLEDGEMENT

The work described in this paper is supported by the Austrian Research Fund, Project No. 4567. I am indebted to R. Gebauer, H. Kredel and F. Winkler for valuable support in the preparation of the examples.

REFERENCES

[6.1] N. K. Bose, *Applied Multidimensional System Theory*, Van Nostrand Reinhold Co., New York, 1982.

[6.2] G. Hermann, 'The Question of Finitely many Steps in Polynomial Ideal Theory (German)', *Mathematische Annalen*, vol. 95, 1926, pp. 736–788.

[6.3] B. Buchberger and R. Loos, 'Algebraic Simplification', in *Computer Algebra – Symbolic and Algebraic Computation*, (B. Buchberger, G. Collins, R. Loos (eds.), Springer, Wien–New York, 1982, pp. 11–43.

[6.4] A. Blass and Yu. Gurevich, 'Equivalence Relations, Invariants, and Normal Forms', *Technical Report*, Dpt. of Math. and Dpt. of Comp. and Commun. Scie., The University of Michigan, Ann Arbor, Michigan.

[6.5] E. Cardoza, R. Lipton, and A. R. Meyer, 'Exponential Space Complete Problems for Petri Nets and Commutative Semigroups', *Conf. Record of the 8th Annual ACM Symp. on Theory of Computing*, 1976, pp. 50–54.

[6.6] E. W. Mayr and A. R. Meyer. 'The Complexity of the Word Problems for Comutative Semigroups and Polynomial Ideals', *Report LCS/TM-199*. M.I.T. Laboratory of Computer Science, 1981.

[6.7] B. Buchberger, 'An Algorithm for Finding a Basis for the Residue Class Ring of a Zero-dimensional Polynomial Ideal (German)', Ph.D. Thesis, Univ. of Innsbruck (Austria), Math. Inst., 1965.

[6.8] B. Buchberger, 'An Algorithmical Criterion for the Solvability of Algebraic Systems of Equations (German)', *Aequationes Mathematicae*, Vol. 4, No. 3, 1970, pp. 374–383.

[6.9] B. Buchberger, 'A Theoretical Basis for the Reduction of Polynomials to Canonical Form', *ACM SIGSAM Bull*. Vol. 10, No. 3, 1976, pp. 19–29.

[6.10] B. Buchberger, 'Some Properties of Gröbner Bases for Polynomial Ideals', *ACM SIGSAM Bull.*, Vol. 10, No. 4, 1976, pp. 19–24.

[6.11] M. Lauer, 'Canonical Representatives for the Residue Classes of a Polynomial Ideal (German)', Diploma Thesis, University of Kaiserslautern (F.R.G.), Dept. of Mathematics, 1976.

[6.12] M. Lauer, 'Canonical Representatives for Residue Classes of a Polyomial Ideal', *Proc. of the 1976 ACM Symp. on Symbolic and Algebraic Computation*, Yorktown Heights, N.Y., August 1976, R. D. Jenks (ed.), pp. 339–345.

[6.13] R. Schrader, 'Contributions to Constructive Ideal Theory (German)', Diploma Thesis, Univ. of Karlsruhe (FRG), Math. Inst., 1976.

[6.14] D. Spear, 'A constructive Approach to Commutative Ring Theory', *Proc. of the MACSYMA Users' Conf.*, Berkeley, July 1977, R. J. Fateman (ed.), published by M.I.T., pp. 369–376.

[6.15] W. Trinks, 'On B. Buchberger's Method for Solving Systems of Algebraic Equations', *J. Number Theory*, Vol. 10, No. 4, 1978, pp. 475–488.

[6.16] F. Winkler, 'Implementation of an Algorithm for Constructing Gröbner Bases (German)', Diploma Thesis, Univ. of Linz (Austria), Dept. of Math., 1978.

[6.17] G. M. Bergman, 'The Diamond Lemma for Ring Theory', *Advances in Math.*, Vol. 29, 1978, pp. 178–218.

[6.18] G. Zacharias, 'Generalized Gröbner Bases in Commutative Polynomial Rings', Bachelor Thesis, M.I.T., Dept. Comp. Scie., 1978.

[6.19] B. Buchberger, 'A Criterion for Detecting Unnecessary Reductions in the Construction of Gröbner Bases', Proc. EUROSAM 79, Marseille, June 1979, W. Ng, (ed.), *Lecture Notes in Computer Science.* Vol. 72, 1979, pp. 3–21.

[6.20] B. Buchberger and F. Winkler, 'Miscellaneous Results on the Construction of Gröbner Bases for Polynomial Ideals I', Techn. Rep. No. 137. University of Linz, Math. Inst., 1979.

[6.21] S. Schaller, 'Algorithmic Aspects of Polynomial Residue Class Rings', Ph.D. Thesis, Techn, Rep. No. 370, Univ. of Wisconsin-Madison, Comp. Scie. Dept., 1979.

[6.22] L. Bachmair and B. Buchberger, 'A Simplified Proof of the Characterization Theorem for Gröbner Bases', *ACM SIGSAM Bull.*, Vol. 14, No. 4, 1980, pp. 29–34.

[6.23] A. M. Ballantyne and D. S. Lankford, 'New Decision Algorithms for Finitely Presented Commutative Semigroups', *Computers and Maths. with Appls.*, Vol. 7, 1981, pp. 159–165.

[6.24] G. Bauer, 'The Representation of Monoids by Confluent Rule Systems', Ph.D. Thesis, University of Kaiserslautern (F.R.G.), Dept. of Comp. Scie., 1981.

[6.25] F. Mora, 'An Algorithm to Compute the Equations of Tangent Cones', *Proc. EUROCAM 82*, Marseille, April 1982, J. Calmet (ed.), Lecture Notes in Comp. Scie., Vol. 144, pp. 158–165.

[6.26] M. Pohst and D. Y. Y. Yun, 'On Solving Systems of Algebraic Equations via Ideal Bases and Elimination Theory', *Proc. of the 1981 ACM Symposium on Symbolic and Algebraic Computation*, Snowbird (Utah), August 1981, P. S. Wang (ed.), published by ACM, pp. 206–211.

[6.27] J. P. Guiver, 'Contributions to Two-dimensional Systems Theory', Ph.D. Thesis, Univ. of Pittsburgh, Math. Dept., 1982.

[6.28] D. Bayer, 'The Division Algorithm and the Hilbert Scheme', Ph.D. Thesis, Harvard University, Cambridge, Mass., Math. Dept., 1982.

[6.29] B. Buchberger, 'A note on the Complexity of Constructing Gröbner bases', *Proc. of the EUROCAL 83*, London, March 1983, H. van Hulzen (ed.), Lecture Notes in Computer Science 162, Springer, 1983, pp. 137–145.

[6.30] B. Buchberger, 'A Critical-pair/completion Algorithm for Finitely Generated Ideals in Rings', *Proc. of the Conf. "Rekursive Kombinatorik"*, Münster, May, 1983, E. Börger, G. Hasenjäger, and D. Rödding (eds.), Lecture Notes in Computer Science 171, Springer, 1983, pp. 137–161.

[6.31] D. Lazard, 'Gröbner Bases, Gaussian Elimination, and Resolution of Systems of Algebraic Equations', *Proc. of the EUROCAL 83*, London, March 1983, H. van Hulzen (ed.), Lecture Notes in Computer Science 162, Springer, 1983, pp. 146–156.

[6.32] R. Llopis de Trias, 'Canonical Forms for Residue Classes of Polynomial Ideals and Term Rewriting Systems', Univ. Aut. de Madrid, Division de Matematicas, submitted to publication, also: Rep. 84–03, Univ. Bolivar, Venezuela.

[6.33] F. Mora and H. M. Möller, 'The Computation of the Hilbert Function', *Proc. of the EUROCAL 83*, London, March, 1983, H. van Hulzen (ed.), Lecture Notes in Computer Science 162, Springer, 1983, pp. 157–167.

[6.34] F. Mora, and H. M. Möller, 'New Constructive Methods in Classical Ideal Theory', Univ. of Genova (Italy), Math. Dept., submitted to publication.

[6.35] A. Galligo, 'The Division Theorem and Stability in Local Analytic Geometry (French)', Extrait des Annales de l'Institut Fourier, Univ. of Grenoble, Vol. 29, No. 2, 1979.

[6.36] H. Hironaka, 'Resolution of Singularities of an Algebraic Variety over a Field of Characteristic Zero: I, II', *Annals of Math.*, Vol. 79, 1964, pp. 109–326.

[6.37] B. Buchberger, 'Miscellaneous Results on Gröber-bases for Polynomial Ideals II', *Techn. Rep.* 83–1, University of Delaware, Dept. of Comp. and Inform. Scie., 1983.

[6.38] D. E. Knuth and P. B. Bendix, 'Simple Word Problems in Universal Algebras', *Proc. of the Conf. on Computational Problems in Abstract Algebra*, Oxford, 1967, J. Leech, (ed.), Pergamon Press, Oxford, 1970.

[6.39] P. Le Chenadec, 'Canonical Forms in Finitely Presented Algebras (French)', Ph.D. Thesis, Univ. of Paris-Sud, Centre d'Orsay, 1983.

[6.40] F. Winkler and B. Buchberger, 'A Criterion for Eliminating Unnecessary Reductions in the Knuth-Bendix Algorithm', *Proc. of the Coll. on Algebra, Combinatorics and Logic in Comp. Scie.*, Györ, Sept. Coll. Math. Soc. J. Bolyai 42, 1985.

[6.41] J. Hsiang, 'Topics in Automated Theorem Proving and Program Generation', Ph.D. Thesis, Univ. of Illinois at Urbana-Champaign, Dept. of Comp. Scie., 1982.

[6.42] B. L. Van der Waerden, *Modern Algebra: I, II*, New York, Frederick Ungar, 1953.

[6.43] L. E. Dickson, 'Finiteness of the Odd Perfect and Primitive Abundant Numbers with n Distinct Prime Factors', *Am. J. of Math.*, Vol. 35, 1913, pp. 413–426.

[6.44] R. Loos, 'Generalized Polynomial Remainder Sequences', in *Computer Algebra – Symbolic and Algebraic Computation*, B. Buchberger, R. Loos, and G. E. Collins, (eds.), Springer, Wien–New York, 2nd edition, 1983, pp. 115–138.

[6.45] B. F. Caviness and R. Fateman, 'Simplification of Radical Expressions', *Proc. 1976 ACM Symposium on Symbolic and Algebraic Computation*, Yorktown Heights, N.Y., August 1976, R. D. Jenks (ed.), published by ACM, pp. 329–338.

[6.46] R. Gebauer and H. Kredel, 'Buchberger's Algorithm for Constructing Canonical Bases (Gröbner bases) for Polynomial Ideals', Program documentation, Univ. of Heidelberg, Dept. for Applied Math., 1983.

[6.47] W. Gröbner, *Modern Algebraic Geometry* (German), Springer, Wien–Innsbruck, 1949.

[6.48] G. E. Collins and L. E. Heindel, 'The SAC–1 Polynomial Real Zero System', *Techn. Rep.* No. 93. Comp. Scie. Dept., Univ. of Wisconsin-Madison, 1970.

[6.49] C. W. Ayoub, 'On Constructing Bases for Ideals in Polynomial Rings over the Integers', *Techn. Rep.* No. 8184, Pennsylvania State Univ., Univ. Park, Dept. of Math., 1981.

[6.50] H. M. Möller, 'Multi-dimensional Hermite Interpolation and Numerical Integration (German)', *Math. Zeitschrift*, Vol. 148, 1976, pp. 107–118.
[6.51] B. Buchberger, V. E. Krishnamurthy, and F. Winkler, 'Gröbner Bases, Polynomial Remainder Sequences and Decoding of Multivariate Hensel Codes', (this volume).

Note added in proof: Meanwhile a number of new papers on Gröbner bases (complexity and applications) have appeared in the literature. Some of them are collected in the following two proceedings [6.52], [6,53]. Some will appear in the new Journal of Symbolic Computation (Academic Press).

[6.52] *Proc. of the EUROSAM 84 Symposium*, Cambridge, J. Fitch (ed.), Springer Lecture Notes in Computer Science 174, 1984.
[6.53] *Proc. of the EUROCAL 85 Symposium*, Linz, April 1985 (to appear.)

Chapter 7

J. P. Guiver

The Equation $Ax = b$ over the Ring $\mathbb{C}[z, w]$

7.1. INTRODUCTION

In Chapter 3 we saw that a strictly causal MIMO system $D_L^{-1}N_L$ was stabilizable by a causal compensator $X_R Y_R^{-1}$ if and only if

$$\det (D_L Y_R + N_L X_R) \in \mathbb{R}_s[z, w].$$

$D_L Y_R + N_L X_R$ represented in some sense the 'denominator' of the feedback system (see (3.92) to (3.95)). It is therefore of interest (see also Emre [7.1]) to study the equation $D_L Y_R + N_L X_R = \Phi$ where Φ is a polynomial matrix and we look for a polynomial solution Y_R, X_R.

This equation can be written

$$A\Xi = \Phi$$

where

$$A = [D_L N_L]$$

and

$$\Xi = \begin{bmatrix} Y_R \\ X_R \end{bmatrix}.$$

In turn, $A\Xi = \Phi$ can be studied as m equations of the form

$$A\xi = \phi_i$$

where ϕ_i is the ith column of Φ and Φ has m columns.

7.2. SUFFICIENT CONDITION FOR SOLUTION

Let

$$A \in \mathbb{C}^{m \times \prime}[z, w] \tag{7.1}$$

and

$$\mathbf{b} \in \mathbb{C}^{m \times 1}[z, w]. \tag{7.2}$$

N. K. Bose (ed.), *Multidimensional Systems Theory*, 233–244

We want to know if there exists $x \in C^{\ell \times 1}[z, w]$ such that

$$Ax = b. \qquad (7.3)$$

We note here that if $A \in \mathbb{R}^{m \times \ell}[z, w]$ and $b \in \mathbb{R}^{m \times 1}[z, w]$ and if $Ax = b$ has a solution with $x \in C^{\ell \times 1}[z, w]$, then by taking the real parts of each side of the equation it follows that (7.3) has a solution in $\mathbb{R}^{\ell \times 1}[z, w]$.

Let D be an integral domain and let K be its ring of quotients. Then for any maximal ideal M of D, its *localization* with respect to M is defined to be the ring

$$D_M = \left\{ \frac{f}{g} \epsilon K: \quad f \epsilon D, \quad g \epsilon D \setminus M \right\}. \qquad (7.4)$$

For the time being we consider a more general situation to (7.3), namely:

$$Ax = b \qquad A \epsilon D^{m \times \ell}, \qquad b \epsilon D^{m \times 1}, \qquad m < \ell. \qquad (7.5)$$

The following result is not difficult to prove (see Gustafson [7.2], Lemma 1).

(7.6) LEMMA. (7.5) has a solution $x \in D^{\ell \times 1}$ if and only if it has a solution over D_M for every maximal ideal M of D.

(7.7) Assume A in (7.5) has full rank when considered as a matrix over K. In other words A has a non-zero m^{th} order minor. Let $I \subset D$ be the ideal generated by the mth order minors of A.

Let

$$I_M \triangleq \left\{ \frac{f}{g} \epsilon K: \quad f \epsilon I, \quad g \notin M \right\}. \qquad (7.8)$$

Then I_M is an ideal in the ring D_M.

(7.9) LEMMA. (7.5) has a solution over D_M if and only if it has a solution over D_M/I_M.

Before proving the preceding lemma, we prove two preliminary results, the first of which appears in (Youla-Gnavi [3.9]), but is proved differently here.

(7.10) LEMMA. *Let A be an $m \times \ell$ matrix with entries in D, $m \leq \ell$.*
Let

$$a_{i_1 \ldots i_m} = A\left(\begin{smallmatrix} 1 & \cdots & m \\ i_1 & \cdots & i_m \end{smallmatrix}\right)^{\dagger}$$

and let I be the ideal of D generated by the $a_{i_1 \ldots i_m}$'s.
*Then, given any $a\epsilon I$, there exists an $\ell \times m$ matrix B with entries in D
such that*

$$AB = aI_m.$$

Proof. For each sequence $1 \leq i_1 < i_2 < \ldots < i_m \leq \ell$ we construct an
$\ell \times m$ matrix $B_{i_1 \ldots i_m}$ as follows:
In rows $i_1, \ldots, i_m$ of $B_{i_1 \ldots i_m}$ we place the m rows of the adjoint of the
matrix $A[\begin{smallmatrix} 1 & \cdots & m \\ i_1 & \cdots & i_m \end{smallmatrix}]^{\dagger}$.
In the remaining rows we place zeroes. It is clear then that

$$AB_{i_1 \ldots i_m} = a_{i_1 \ldots i_m} I_m.$$

Now if $a\epsilon I$ then $\exists c_{i_1 \ldots i_m} \epsilon D$ for each sequence $1 \leq i_1 < i_2 < \ldots <
i_m \leq \ell$ such that

$$a = \sum_{1 \leq i_1 < \ldots < i_m \leq \ell} c_{i_1 \ldots i_m} a_{i_1 \ldots i_m}.$$

Let

$$B = \sum_{1 \leq i_1 < \ldots < i_m \leq \ell} c_{i_1 \ldots i_m} B_{i_1 \ldots i_m}.$$

Then

$$AB = A\left(\sum c_{i_1 \ldots i_m} B_{i_1 \ldots i_m} \right)$$

$$= \sum c_{i_1 \ldots i_m} AB_{i_1 \ldots i_m}$$

$$= \sum c_{i_1 \ldots i_m} a_{i_1 \ldots i_m} I_m$$

$$= a\, I_m.$$

$\dagger$ $A[\begin{smallmatrix} i_1 & \cdots & i_m \\ j_1 & \cdots & j_m \end{smallmatrix}]$ denotes the submatrix of a matrix A obtained by taking rows i_1,
$\ldots, i_m$ and columns $j_1, \ldots, j_m$
$A(\begin{smallmatrix} i_1 & \cdots & i_m \\ j_1 & \cdots & j_m \end{smallmatrix})$ denotes the determinant of $A[\begin{smallmatrix} i_1 & \cdots & i_m \\ j_1 & \cdots & j_m \end{smallmatrix}]$

(7.11) COROLLARY. *Let A, D and I be as in Lemma (7.10). Then if ϕ is any $m \times k$ matrix with entries in I there exists an $\ell \times k$ matrix B with entries in D such that*

$$AB = \Phi.$$

Proof. Let E_{ij} denote the $m \times k$ matrix with zeroes everywhere except in the (i, j) position where there is a one.

Let B_{ij} be an $\ell \times m$ matrix with entries in D such that

$$AB_{ij} = \phi_{ij}I_m$$

where ϕ_{ij} is the $(i, j)^{th}$ entry of Φ.

Then

$$AB_{ij}E_{ij} = \phi_{ij}E_{ij}$$

Let

$$B = \sum_{i, j} B_{ij}E_{ij}$$

Then

$$AB = A(\sum_{i, j} B_{ij}E_{ij})$$

$$= \sum_{i, j} AB_{ij}E_{ij}$$

$$= \sum \phi_{ij}E_{ij} = \Phi.$$

Proof of Lemma (7.9): Suppose (7.5) has a solution over D_M/I_M then there exists $\mathbf{x} \in D_M^{\ell \times 1}$ and $\mathbf{y} \in I_M^{m \times 1}$ such that

$$A\mathbf{x} = \mathbf{b} + \mathbf{y}.$$

Multiplying both sides by a common denominator of the $\mathbf{x}$ and $\mathbf{y}$ gives us

$$A\mathbf{x}^* = g\mathbf{b} + \mathbf{y}^*$$

where

$$\mathbf{x}^* \in D^{\ell \times 1}, \quad \mathbf{y}^* \in I^{m \times 1} \quad \text{and} \quad g \in D \setminus M.$$

Now from Corollary (7.11), there exists $\tilde{\mathbf{x}} \epsilon D^{\ell \times 1}$ such that

$$A\tilde{\mathbf{x}} = \mathbf{y}^*.$$

Therefore

$$A(\mathbf{x}^* - \tilde{\mathbf{x}}) = g\mathbf{b}$$

where

$$\mathbf{x}^* - \tilde{\mathbf{x}} \epsilon D^{\ell \times 1} \quad \text{and} \quad g \notin M$$

So

$$A \frac{\mathbf{x}^* - \tilde{\mathbf{x}}}{g} = \mathbf{b} \quad \text{and} \quad \frac{\mathbf{x}^* - \tilde{\mathbf{x}}}{g} \epsilon D_M^{\ell \times 1}.$$

The proof the other way is trivial.

Now let us return to Equation (7.3), i.e. when $D = \mathbb{C}[z, w]$. We assume, in addition, that

(7.12) the mth order minors are relatively prime.† This in particular implies that they have only finitely many zeroes in common. Denote the (non-zero) m^{th} order minors by $f_1, \ldots f_r$ and let the common zeroes of $f_1, \ldots f_r$ be $\mathbf{p}_1, \ldots \mathbf{p}_k$ where

$$\mathbf{p}_i = (a_i, b_i) \epsilon \mathbb{C}^2 \quad i = 1, \ldots k.$$

Now it is well known (see for example Fulton [7.3]) that the maximal ideals of $\mathbb{C}[z, w]$ are precisely those of the form $M = \langle z - a, w - b \rangle$ for some $(a, b) \epsilon \mathbb{C}^2$. Suppose $\mathbf{p} = (a, b)$ is not one of the $\mathbf{p}_i$'s. Form the polynomial

$$h(z, w) = \prod_{i=1}^{s} (z - a_{\mu_i}) \prod_{j=1}^{t} (w - b_{\nu_j}) \tag{7.13}$$

where

$$\{\mu_1, \ldots, \mu_s, \nu_1, \ldots, \nu_t\} = \{1, \ldots k\}$$

and

$$a_{\mu_i} \neq a \quad i = 1, \ldots s$$
$$b_{\nu_i} \neq b \quad j = 1, \ldots t.$$

† This assumption holds when $A = [D_\ell \ N_\ell]$ with $D_\ell^{-1}N_\ell$ irreducible.

Then $h(a, b) \neq 0$ but $h(a_i, b_i) = 0 \qquad i = 1, \ldots k.$
Therefore

$$h^d \epsilon I \text{ for some positive integer } d$$

by Hilbert's Nullstellensatz. So we have that

$$h^d \mathbf{b} \epsilon I^{m \times 1}$$

and hence by Corollary (7.11) there exists

$$\mathbf{x} \epsilon \mathbb{C}^{\ell \times 1}[z, w] \text{ such that}$$

$$A\mathbf{x} = h^d \mathbf{b}$$

Therefore $A\mathbf{x}/h^d = \mathbf{b}$ is a solution of (7.3) over $\mathbb{C}[z, w]_M$ (since $h(a, b) \neq 0$).

Therefore (7.3) has a solution over $\mathbb{C}[z, w]_M$ for every $M = \langle z - a, w - b \rangle$ such that (a, b) is not one of the $\mathbf{p}_i$'s.

(7.14) NOTATION. We let M_i denote the maximal ideal $\langle z - a_i, w - b_i \rangle$ of $\mathbb{C}[z, w]$. Let D_i denote the localization of $\mathbb{C}[z, w]$ with respect to M_i, and let $I_i = I_{M_i}$ (see 7.8).

Then the above discussion together with Lemmas (7.6) and (7.9) imply the following lemma.

(7.15) LEMMA. Equation (7.3), where A satisfies (7.7) and (7.12), has a solution over $\mathbb{C}[z, w]$ if and only if it has a solution over D_i/I_i for each $i = 1, \ldots k.$

$\mathbb{C}$ is embedded naturally in D_i/I_i and D_i/I_i is a vector space over $\mathbb{C}$ (as well as having a ring structure) which is finite dimensional (see Appendix A).

The easiest case to deal with is when $D_i/I_i \cong \mathbb{C}, i = 1, \ldots k$; here the isomorphism is given by

$$\frac{f}{g} + I_i \longrightarrow \frac{f(a_i, b_i)}{g(a_i, b_i)} \tag{7.16}$$

and the condition $D_i/I_i \cong \mathbb{C}, i = 1, \ldots k$ is equivalent to $I = \text{Rad}(I)$ (see Appendix A).

(7.15) and (7.16) therefore lead to the following condition.

(7.17) If $I = \text{Rad}(I)$, then (7.3) (with A satisfying (7.7) and (7.12)) has a solution over $\mathbb{C}[z, w]$ if and only if

Rank A = Rank $[Ab]$ when evaluated at (a_i, b_i)

$i = 1, \ldots k.$

In general D_i/I_i is not a field but is a Noetherian full quotient ring.

(7.18) DEFINITION. (see Ching [7.4]) A commutative ring with identity is said to be a Noetherian full quotient ring if it is Noetherian with the property that every element is either a unit or a zero divisor.

(7.19) LEMMA. *D_i/I_i is a Noetherian full quotient ring.*
   Proof. If $f/g + I_i \epsilon D_i/I_i$, $g(a_i, b_i) \neq 0$, is such that $f(a_i, b_i) \neq 0$, then $g/f + I_i$ is the inverse and $f/g + I_i$ is a unit.

Suppose $f/g + I_i$, $g(a_i, b_i) \neq 0$,   is such that $f(a_i, b_i) = 0.$

Let s be the smallest integer such that there exists $q \epsilon \mathbb{C}[z, w]$ with $q(a_i, b_i) \neq 0$ and $qf^s \epsilon I$. Certainly a finite minimum s does exist since if we choose h such $h(a_j, b_j) = 0, j \neq i, h(a_i, b_i) \neq 0$ (in a similar manner to the construction in (7.13)) then by Hilbert's Nullstellensatz $h^d f^d \epsilon I$ for some d.
   Let q be such that $qf^s \epsilon I$.
Then

$$(qf^{s-1} + I_i)\left(\frac{f}{g} + I_i\right) = \frac{qf^s}{g}\epsilon I_i \ (s > 0 \quad \text{since} \quad 1 \notin I_i).$$

But qf^{s-1} is not in I_i, for if it were, $qf^{s-1} = q_1/q_2$ where $q_2(a_i, b_i) \neq 0$ and $q_1 \epsilon I$; therefore $q_2 qf^{s-1} \epsilon I$ contradicting the minimality of s.
   So $qf^{s-1} + I_i$ is a non-zero element of R_i/I_i annihilating $f/g + I_i$, and $f/g + I_i$ is a zero divisor.
   Finally we observe that R_i/I_i is Noetherian since the Noetherian property is closed under the operations of localization and formation or factor rings.
   An important property of Noetherian full quotient rings is the following:

(7.20) THEOREM. (see Ching [7.4], Kaplansky [7.5, Theorem 82]). A finite set of elements in a Noetherian full quotient ring either contains only zero divisors in which case these have a common annihilator, or contains a unit and therefore generates the whole ring.

Ching [7.4, Th. 6] presents a sufficient condition for solution of a linear equation over a Noetherian full quotient ring in terms of McCoy rank.

(7.21) DEFINITION. Let R be a commutative ring with identity. Let A be a matrix with entries in R. The McCoy rank of A denoted by $\text{rank}_M A$ is the greatest integer r such that the rth order minors of A do not have a common annihilator. $\text{Rank}_M A$ is defined to be zero if all the entries have a common annihilator.

If R is a Noetherian full quotient ring, Theorem (7.20) allows us to describe McCoy rank in a simpler manner; namely $\text{rank}_M A$ is the maximum integer r such that there is an rth order minor of A which is a unit. We therefore have:

(7.22) For $A \in \mathbb{C}^{m \times \ell}[z, w]$, $\text{rank}_M A$ over the ring D_i/I_i is just the usual rank of A over $\mathbb{C}$ when A is evaluated at (a_i, b_i). We let ρ_i denote the usual rank of A at (a_i, b_i), $i = 1, \ldots k$.

Ching's condition, mentioned above, is as follows:
(7.23) Let $A \in R^{m \times \ell}$, $\mathbf{b} \in R^{m \times 1}$ where R is a Noetherian full quotient ring and A is not of full McCoy rank over R. Then $A\mathbf{x} = \mathbf{b}$ has a solution over R if those $(\text{rank}_M A + 1)^{th}$ order minors of the matrix $[A\mathbf{b}]$ which are not minors of A (i.e. which involve the column $\mathbf{b}$) are all zero in R.

Note that if A has full McCoy rank over R then it has an mth order minor which is a unit in R (when R is Noetherian full quotient) and therefore $A\mathbf{x} = \mathbf{b}$ will have a solution in R.

As an immediate consequence we get the following theorem.

(7.24) THEOREM. Let $A \in \mathbb{C}^{m \times \ell}[z, w]$ satisfy (7.7) and (7.12). Let ρ_i $i = 1, \ldots k$ be as in (7.22). Then a sufficient condition for (7.3) to have a solution in $\mathbb{C}^{\ell \times 1}[z, w]$ is that for $i = 1, \ldots k$, each $(\rho_i + 1)^{th}$ order minor of $[A\mathbf{b}]$ involving $\mathbf{b}$ lies in I_i. See appendix A for the problem of determining when a given polynomial is in I_i.

(7.25) COROLLARY. *Let A, ρ_i be as in the above theorem. Let $\rho = min\langle \rho_1,$ $\ldots, \rho_k \rangle$. Then (7.3) has a solution if each $(\rho + 1)^{th}$ order minor of $[A\mathbf{b}]$ involving $\mathbf{b}$ lies in I.*

Proof. The $(\rho + 1)^{th}$ order minors involving $\mathbf{b}$ will in particular lie in I_i (in fact they lie in I_i for $i = 1, \ldots k$, if and only if they lie in I – see corollary A2). Therefore so will the $(\rho_i + 1)^{\text{th}}$ order minors involving $\mathbf{b}$. Now use the previous theorem.

APPENDIX A. ZERO-DIMENSIONAL POLYNOMIAL IDEALS

We recall in this appendix a few facts from the theory of polynomial ideals (see for example Van der Waerden [7.6] and Fulton [7.3]). For simplicity we work over $\mathbb{C}$ although any algebraically closed field would do.

Let I be an ideal of the polynomial ring $\mathbb{C}[z_1, \ldots, z_n]$. We let $V(I)$ denote the set $\{(a_1, \ldots, a_n) \epsilon \mathbb{C}^n : f(a_1, \ldots, a_n) = 0 \; f \epsilon I\}$. $V(I)$ is called the zero set of I. Hilbert's basis theorem tells us that any polynomial ideal is finitely generated, so $V(I)$ can be characterized as the set of common zeroes of a finite set of generating polynomials of I. Given $f_1, \ldots, f_r \epsilon \mathbb{C}[z_1, \ldots, z_n]$, we denote by $\langle f_1, \ldots, f_r \rangle$ the ideal in $\mathbb{C}[z_1, \ldots, z_n]$ generated by the f_i's.

Given an ideal I in any ring R, the radical of I is defined as $\mathrm{Rad}(I) = \{f \epsilon R : f^n \epsilon I \text{ for some positive integer } n\}$. An ideal is called a radical ideal if $I = \mathrm{Rad}(I)$.

Given a set $X \subset \mathbb{C}^n$ we denote by $I(X)$ the ideal of polynomials in $\mathbb{C}[z_1, \ldots, z_n]$ which are zero at each point of X. Clearly $I(X)$ is a radical ideal for any $X \subset \mathbb{C}^n$. If I is any ideal of $\mathbb{C}[z_1, \ldots, z_n]$ it is also clear that $\mathrm{Rad}(I) \subset I(V(I))$. Hilbert's Nullstellensatz gives us the inclusion the other way, which then implies that $\mathrm{Rad}(I) = I(V(I))$; the fact that $\mathbb{C}$ is algebraically closed is essential here.

In fact, given a polynomial ideal $I = \langle f_1, \ldots, f_r \rangle \subset \mathbb{C}[z_1, \ldots, z_n]$, there exists a number d such that if $f \epsilon I(V(I))$, then $f^d \epsilon I$; in other words the exponent d depends only on I and not on the particular polynomial $f \epsilon I(V(I))$.

Closely related to Hilbert's Nullstellensatz is the fact that the maximal ideals of $\mathbb{C}[z_1, \ldots, z_n]$ are precisely those ideals of the form $M = \langle z_1 - a_1, z_2 - a_2, \ldots, z_n - a_n \rangle$ *where* $a_i \epsilon \mathbb{C}$, $i = 1, \ldots, n$. Given such an ideal we can define the localization of $\mathbb{C}[z_1, \ldots, z_n]$ with respect to M to be the ring $D_M = \{f/g \epsilon \mathbb{C}(z_1, \ldots, z_n) : g(a_1, \ldots, a_n) = 0\}$. Let I_M denote the ideal of D_M defined by $I_M = \{f/g \epsilon \mathbb{C}(z_1, \ldots, z_n) : g(a_1, \ldots, a_n) \neq 0, f \epsilon I\}$.

From now on until the end of the appendix we will assume I is an ideal of $\mathbb{C}[z_1, \ldots, z_n]$ such that $V(I)$ contains only a finite non-zero number of points; such an ideal is called zero-dimensional. Label these points $\mathbf{p}_1, \ldots, \mathbf{p}_k$ where $\mathbf{p}_i = (a_1^{(i)}, \ldots, a_n^{(i)})$, $i = 1, \ldots, k$. Let M_i denote the maximal ideal $\langle z_1 - a_1^{(i)}, \ldots, z_n - a_n^{(i)} \rangle$ and let D_i and I_i denote D_{M_i} and

I_{M_i} respectively, $i = 1, \ldots, k$. Then Fulton [7.3, Chapter 2, Proposition 6] presents the following result.

(A1) PROPOSITION. There is a natural isomorphism from $\mathbb{C}[z_1, \ldots, z_n]/I$ onto the direct product $X_{i=1}^k D_i/I_i$.

(A2) COROLLARY. $f \epsilon I$ if and only if $f \epsilon I_i$ $i = 1, \ldots k$.

The field $\mathbb{C}$ is naturally embedded in D_i/I_i and D_i/I_i is in fact finite dimensional as a vector space over $\mathbb{C}$ because (see Fulton [7.3 Chapter 1, corollary 4]) $\mathbb{C}[z_1, \ldots, z_n]/I$ is finite dimensional when $V(I)$ is finite. For each i, $i = 1, \ldots k$, there exists a minimum positive integer d_i such that $h \epsilon I_i$ for any monomial h in $z_1 - a_1^{(i)}, \ldots, z_n - a_n^{(i)}$ of degree d_i or greater. This follows from Hilbert's Nullstellensatz; for let $f \epsilon \mathbb{C}[z_1, \ldots, z_n]$ be such that $f(\mathbf{p}_j) = 0, j \neq i, f(\mathbf{p}_i) \neq 0$ (see for example the construction in (7.13)), then $(z_j - a_j^{(i)})^d f^d \epsilon I$, where d is the exponent in Hilbert's Nullstellensatz, and therefore $(z_j - a_j^{(i)})^d \epsilon I_i$ $j = 1, \ldots, n$.

Consequently any monomial in the $(z_j - a_j^{(i)})$'s of degree greater than $n(d - 1)$ is in I_i which gives us an upper bound for d_i. D_i/I_i as a vector space is then a subspace of the space of polynomials in $z_1 - a_1^{(i)}, \ldots, z_n - a_n^{(i)}$ (strictly speaking residue classes of polynomials mod I_i) of degree less than d_i. The simplest case is when $d_i = 1$ in which case D_i/I_i is isomorphic to $\mathbb{C}$ because every element $f/g + I_i \epsilon D_i/I_i$ is equal to a constant mod I_i; the isomorphism is given by $f/g + I_i \rightarrow [f(\mathbf{p}_i)]/[g(\mathbf{p}_i)]$ because if $c \epsilon \mathbb{C}$ is such that $f/g - c \epsilon I_i$ then $f/g - c$ evaluated at $\mathbf{p}_i$ must be zero. So if $d_i = 1$, $(f/g) \epsilon I_i$ if and only if $f(\mathbf{p}_i) = 0$. Consequently, if $d_i = 1$, $i = 1, \ldots, k$, then, for $f \epsilon \mathbb{C}[z_1, \ldots, z_n]$, $f \epsilon I$ if and only if $f \epsilon I_i$ for each i which is true if and only if $f(\mathbf{p}_i) = 0$ for each i. Therefore if $d_i = 1$, $i = 1, \ldots, k$, then $\mathrm{Rad} I = I$; it is also clear that the converse is true since if $\mathrm{Rad} I = I$ it is easy to see that for each i, j, $i = 1, \ldots k, j = 1, \ldots n$, $(z_j - a_j^{(i)}) \epsilon I_i$ and therefore $d_i = 1$.

A slightly different approach makes use of the primary decomposition theorem for ideals in Noetherian rings. In particular, if $I \subset \mathbb{C}[z_1, \ldots, z_n]$ is such that $V(I)$ is finite then $I = Q_1 \cap Q_2 \ldots \cap Q_k$ where Q_i is the ideal $M_i^{\sigma_i} + I$, M_i is the maximal ideal $\langle z_1 - a_1^{(i)}, \ldots, z_n - a_n^{(i)} \rangle$ and σ_i is the smallest number σ such that $M_i^\sigma \subset M_i^{\sigma+1} + I_i$ (see van der Waerden [7.6, § 96] for this and for what follows. The condition that a given polynomial f is in I if (and only if) $f \epsilon M_i^{\sigma_i} + I_i$ for $i = 1, \ldots, k$ is known as the fundamental theorem of Max Noether. It is not difficult to show that σ_i is the same as the d_i discussed earlier in the appendix. It is also possible to construct polynomials $h_1, \ldots, h_k$ in $z_1 - a_1^{(i)}, \ldots, z_n - a_n^{(i)}$ each of whose degrees is less than σ_i such that $f \epsilon Q_i$ if and only if f, when expanded in a Taylor series about the point $(a_1^{(i)}, \ldots, a_n^{(i)})$ and after omitting all terms of degree $\geq \sigma_i$, is equal to a linear combination of $h_1, \ldots, h_k$ with constant coefficients. This is also equivalent to having a basis for those in $z_1 - a_1^{(i)}, \ldots, z_n - a_n^{(i)}$ of degree less than $d_i (= \sigma_i)$ which belong to I_i; in fact $Q_i = I_i \cap \mathbb{C}[z_1, \ldots, z_n]$ – this follows, for example, from [7.7, § 2.6 Cor. 2] and the easily seen fact that $D_i Q_i = I_i$, although it is not difficult to prove from first principles. So we can actually establish whether or not a given polynomial is in Q_1 (or equivalently I_i).

The simplest case that arises is when $\sigma_i = 1 (= d_i)$, $i = 1, \ldots, k$. As we have seen earlier this occurs if and only if $D_i / I_i \cong \mathbb{C}$ $i = 1, \ldots, k$ which in turn is equivalent to $\mathrm{Rad} I = I$. This case is also considered by Van der Waerden as an application of Noether's theorem. He shows that $\sigma_i = 1, i = 1, \ldots, k$ occurs if the linear terms of $f_1, \ldots f_r$ when expanded in a Taylor Series about $\mathbf{p}_i$ span the whole n dimensional space of linear terms in $z_1 - a_1^{(i)}, \ldots, z_n - a_n^{(i)}$; the converse of this statement is also seen to be true because $\sigma_i = 1 \Rightarrow z_j - a_j^{(i)} \epsilon I_i$ $j = 1, \ldots, n \Rightarrow (z_j - a_j^{(i)})g = \Sigma_{s=1}^{r} g_s f_s$ for some g, $g_s \epsilon \mathbb{C}[z_1, \ldots, z_n]$ with $g(\mathbf{p}_i) \neq 0$. Consequently we have the following result:

(A3) PROPOSITION. Let $I = \langle f_1, \ldots, f_r \rangle$, $\mathbf{p}_i = (a_1^{(i)}, \ldots, a_n^{(i)})$ be as above.
   Then the following are equivalent:
   (i) $D_i / I_i \cong \mathbb{C}$ $i = 1, \ldots, k$
   (ii) $\mathrm{Rad} I = I$
   (iii) $J(a_1^{(i)}, \ldots, a_n^{(i)})$ is of rank n, $i = 1, \ldots, k$
   where $J(z_1, \ldots, z_n)$ is the matrix

$$\begin{bmatrix} \dfrac{\partial f_1}{\partial z_1} \cdots\cdots\cdots \dfrac{\partial f_1}{\partial z_n} \\[2em] \dfrac{\partial f_r}{\partial z_1} \cdots\cdots\cdots \dfrac{\partial f_r}{\partial z_n} \end{bmatrix}$$

We also have in the above proposition that (i) is equivalent to (iii) for each i separately.

REFERENCES

[7.1] E. Emre, 'The Polynomial Equation $QQ_c + RP_c = \phi$ with Application to Dynamic Feedback', *SIAM J. Control and Optimization*, Vol. 18, No. 6, November 1980, pp. 611–620.

[7.2] W. Gustafson, 'Roth's Theorem over Commutative Rings', *Linear Algebra and its Applications*, 23, 1979, pp. 245–251.

[7.3] W. Fulton, *Algebraic Curves: An Introduction to Algebraic Geometry*, Benjamin/Cummings, Massachusetts, 1969.

[7.4] W. S. Ching, 'Linear Equations over Commutative Rings', *Linear Algebra and its Applications*, 18, 1977, pp. 257–266.

[7.5] I. Kaplansky, *Commutative Rings*, Allyn and Bacon, Boston, 1970.

[7.6] B. van der Waerden, *Modern Algebra*, Vol. II, Ungar, NY, 1950.

[7.7] D. Northcott, *Ideal Theory*, Cambridge University Press, 1953.

Chapter 8

Open Problems

REMARKS BY N. K. BOSE

The open problems in this chapter characterize the variety of mathematical resources that are required for investigations into their solutions.
Besides being theoretically challenging, additional motivation is derived
from the fact that the solutions to these problems will have significant
applications in multidimensional systems analysis and design. The names
of the contributors to this chapter on open problems, other than those
contributed by me, are explicitly mentioned at relevant places.

PROBLEM # 1

*Problem on BIBO Stability With Presence of Non-Essential Singularities
of the Second Kind of T^n*

Let $H(\mathbf{z}) = [P(\mathbf{z})]/[Q(\mathbf{z})]$ be a n-variate rational function where $P(\mathbf{z})$ and
$Q(\mathbf{z})$ are relatively prime polynomials having real coefficients and
$Q(\mathbf{0}) \neq 0$. Then, $H(\mathbf{z})$ is expandable as a power series around $\mathbf{z} = \mathbf{0}$.

$$H(\mathbf{z}) = \sum_{k_1 = 0}^{\infty} \cdots \sum_{k_n = 0}^{\infty} h(\mathbf{k}) z_1^{k_1} \cdots z_n^{k_n}.$$

Clearly $\mathbf{k} \triangleq (k_1, \ldots, k_n)$ and $\mathbf{z} \triangleq (z_1, \ldots, z_n)$.
Investigate the truth or falsity of the following conjectures.
   Conjecture 1. Let there be a finite number of points[†] on T^n where $P(\mathbf{z})$
and $Q(\mathbf{z})$ have common zeros. Then, the series $\{h(\mathbf{k})\}$ is absolutely
summable if and only if $H(\mathbf{z})$ is continuously extendable from U^n to $\bar{U}^n$.
   Conjecture 2. A BIBO stable rational $H(\mathbf{z})$ with nonessential singularities of the second kind on T^n cannot have a BIBO stable inverse.

PROBLEM # 2

Finite Length Sequence Reconstruction from Specified Phase Samples

The importance of phase has been recognized in many applications
including digital image processing. It has been confirmed that the fidelity
of reconstruction of an object from the available magnitude only of its
Fourier transform often stops tantalizingly short of that which would
allow a recovery of the object from the available phase information only

---

[†] Also the case when this restriction is not satisfied remains to be investigated.

N. K. Bose (ed.), Multidimensional Systems Theory, 245–260.
© *1985 by D. Reidel Publishing Company.*

of its Fourier transform. With so much information contained in the phase, it is natural to attempt classification of signals which can be recovered completely or up to, possibly, a scale factor from the phase specifications. Minimum-phase signals are known to form one such distinguishing class. Oppenheim and Lim [1] proved the following result.

Fact. Let $(h_0, h_1, \ldots, h_{N-1})$ be a finite length sequence of length N. Associate, the generating function or z-transform,

$$H(z) = \sum_{k=0}^{N-1} h_k z^{-k}$$

with the sequence. The sequence is uniquely specified to within a scale factor by $(N-1)$ distinct samples of its phase in the interval $0 < \omega < \Pi$ if $H(z)$ has no zeros either on $|z| = 1$ or in conjugate reciprocal pairs, i.e. $h_k \neq h_{N-k-1}^*, k = 0, 1, \ldots, N-1$ or $H(z) \neq \pm z^{-(N-1)} H^*(z^{-1})$.

Hayes obtained a multidimensional counterpart [2] of the above result. Specifically, he showed that a n-dimensional sequence which has a finite extent lattice (grid) support of $N_1 \times N_2 \times \ldots N_n$ grid points is uniquely specified to within a scale factor by the sampled phase (required in the Discrete Fourier Transform or DFT implementation) over at least $(2N_1 - 1) \times (2N_2 - 1) \times \ldots \times (2N_n - 1)$ points, provided the associated z transform, $H(\mathbf{z})$ is 'nonsymmtric' i.e. $H(\mathbf{z}) \neq \pm z_1^{-k_1} z_2^{-k_2} \ldots z_n^{-k_n} H^*(\mathbf{z}^{-1})$ for any n-tuple $(k_1, k_2, \ldots, k_n)$ of positive integers $k_1, k_2, \ldots, k_n$. It is required to investigate how these results can be generalized to a broader class of sequences with infinite support, whose z-transforms may or may not be rational. Also, acknowledging the discussions with Mr. P. Liang the results in [3] and [4] might provide some helpful background for investigation into the theoretically challenging problem of determining the feasibility of reducing the number of phase samples (at a symmetrically arrayed set of points on the unit circle in the 1–D case or on the distinguished boundary of the polydisc in the n–D case starting from $z_i = e^{j\theta_i}$, where one or more θ_i, $i = 1, 2, \ldots, n$, may not be zero) necessary for reconstructing the complete sequence. In that event, the fast algorithms available for implementing the DFT might not be applicable though the reduced number of samples might have other advantages with regard to the overall computational complexity.

REFERENCES

[1] A. V. Oppenheim and J. S. Lim, 'The Importance of Phase in Signals', *Proc. of IEEE*, 69, May 1981, pp. 529–541.

[2] M. H. Hayes, 'The Reconstruction of a Multidimensional Sequence from the Phase or Magnitude of its Fourier Transform', *IEEE Trans. ASSP*, 30, April 1982, pp. 140–154.
[3] J. L. Schiff and W. J. Walker, 'Finite Harmonic Interpolation', *J. Math. Analysis and Applications*, 86, 1982, pp. 648–658.
[4] J. L. Schiff and W. J. Walker, 'Finite Harmonic Interpolation, 2', *J. Math. Analysis and Applications*, 87, 1982, pp. 1–8.

PROBLEM # 3

Conditions for Irreducibility of Multivariate Polynomials Over a Real Field

Let $F(\omega_1, \omega_2, \ldots, \omega_n) \leftrightarrow f(x_1, x_2, \ldots, x_n)$ be $n - D$ Fourier transform pairs. Denoting $(\omega_1, \omega_2, \ldots, \omega_n)$, $(x_1, x_2, \ldots, x_n)$ by $\boldsymbol{\omega}, \mathbf{x}$, respectively, let the magnitude and phase of $F(\boldsymbol{\omega})$, $f(\mathbf{x})$, be described below.

$$F(\boldsymbol{\omega}) = |F(\boldsymbol{\omega})| \exp j\phi(\boldsymbol{\omega})$$

$$f(\mathbf{x}) = |f(\mathbf{x})| \exp \ j\theta(\mathbf{x}).$$

When $n = 1$, the problem of recovering *phase* $\phi(\boldsymbol{\omega})$ from a specified magnitude $|F(\omega)|$ occurs in network theory. Unique phase recovery is ensured through the requirement of a zero-free half plane for the analytic function $F(\omega)$ [1]. In electrical engineering terminology, this constraint is referred to as the minimum (maximum) phase constraint [2]. The $n = 2$ case becomes, especially, important in problems of electron microscopy, wavefront sensing, X-ray diffraction, astronomy, crystallography and imaging by speckle interferometry. Fienup [3] provides a summary of iterative algorithms available for retrieval of $\phi(\omega)$ from $|F(\omega)|$, along with analysis of other variations of this basic problem where constraints (like nonnegativity) might be imposed depending on the nature of applications. Manolitsakis [4] formulates a general description of n-dimensional Fourier transforms in terms of the zero surfaces of functions of several complex variables, and applies this description to the $n = 2$ case to show that ambiguity (or nonuniqueness) of object reconstruction (or phase) from intensity (or magnitude) data depends on the number, N, of irreducible factors, $F_m(\boldsymbol{\omega})$, into which an entire function $F(\boldsymbol{\omega})$ is decomposable.

$$F(\boldsymbol{\omega}) = \prod_{m=1}^{N} [F_m(\boldsymbol{\omega}) \ e^{\gamma_m}]^{\ell_m}, \qquad (N \le \infty)$$

In the above equation, ℓ_m's are integers, γ_m's are polynomials in ω and

$F_m(\omega)$'s are irreducible entire functions. Fiddy, Brames and Dainty [5] consider the construction of irreducible bivariate polynomials so that the uniqueness of phase is guaranteed if a model for the object is based on the condition that its Fourier transform is an irreducible polynomial. They provide a sufficient condition for a bivariate polynomial to be irreducible. *It will be useful to obtain necessary and/or sufficient conditions for a specified multivariate polynomial to be irreducible over the real field of coefficients.*

REFERENCES

[1] R. E. Burge, M. A. Fiddy, A. H. Greenaway, and G. Ross, 'The Phase Problem', *Proc. R. Soc. Lond. A.* 350, 1976, pp. 191–212.

[2] N. Balabanian and T. A. Bickart, '*Linear Network Theory: Analysis, Properties, Design and Synthesis*', Matrix Publishers, Beaverton, OR, 1981.

[3] J. R. Fienup, 'Phase Retrieval Algorithms: A Comparison', *Applied Optics*, 21, 1982, pp. 2758–2769.

[4] I. Manolitsakis, 'Two-dimensional Scattered Fields: A Description in Terms of the Zeros of Entire Functions', *J. Math. Phys.*, 23, Dec. 1982, pp. 2291–2297.

[5] M. A. Fiddy, B. J. Brames, and J. C. Dainty, 'Enforcing Irreducibility for Phase Retrieval in Two Dimensions', *Optics Letters*, 8, Feb. 1983, pp. 96–98.

PROBLEM # 4

Conditions for Feasibility of Multidimensional Spectrum Estimation

Let $r_0, r_1, \ldots, r_k$ be the estimated lags of the autocorrelation function of a time series. The autocorrelation matrix, A, which is symmetric Toeplitz, has a typical element,

$$a_{ij} = r_{|i-j|}, \qquad 1 \le i, j \le k + 1$$

in its ith row and jth column. It is well known that the given autocorrelation lags can be uniquely extended to an autocorrelation function defined over all integers if and only if the matrix A is positive definite. This fact provides useful support to justify the 1–D spectrum estimation methods. The n–D spectral estimation problem requires the estimation of the power spectrum $S(\omega)$ of a stationary and homogeneous zero-mean random field $u(\mathbf{x})$, specified by its autocorrelation function,

$$r(\delta) = E[u(\mathbf{x})u^*(\mathbf{x} + \delta)]$$

where $\delta \epsilon \Delta = \{0, \pm\delta_1, \ldots, \pm\delta_L\}$. Alternatively, the question (*extend-*

ability question) posed is: for a finite set of correlation samples $\{r_0(\delta):$ $\delta\epsilon\Delta\}$, does there exist a positive spectrum $S(\omega)$ such that for $\delta\epsilon\Delta$, where Δ is a finite index set symmetric about the origin and including it,

$$r(\delta) = \int S(\omega) \, e^{j<\omega,\,\delta>} \, d\omega, \text{ with } r(\delta) = r_0(\delta), \delta\epsilon\Delta$$

where the domain of integration is the domain of support of $S(\omega)$? Dickinson [1] pointed out the constraint of homogeneity is so strong that not all sets of homogeneous random variables on a square are extendable to homogeneous random fields over all $Z^2 = Z \times Z$. Therefore, the requirement of positive-definiteness of the block Toeplitz correlation matrix is not sufficient for a finite set of correlation samples to be extendable in the n–D case, $n \geqslant 2$.

The reason for this can be viewed in terms of certain fundamental limitations that exist when trying to extend 1–D results to 2–D situations. Rudin [1.55] used a result of Hilbert (if $N \geq 3$, there are polynomials $F(x_1, x_2)$ of degree $2N$ which are positive for all real x_1, x_2 and which are not sums of squares of polynomials) to derive an analogous result for trigonometric polynomials on the 2–D torus, T^2. Furthermore, for every $M \geq 3$, there are functions which are positive definite on the square $S_M = \{(k_1, k_2)\epsilon Z^2, 0 \leq k_1, k_2 \leq M\}$, but which cannot be extended in any way to be positive definite on Z^2. This fact can be interpreted in terms of an established result that positive linear maps between C^*-algebra of operators with an identity are not the natural generalization of positive linear functionals. [1.48], [2]. McClellan [1.46] has reviewed results available on functions that are extendable in the context of multidimensional spectrum estimations. The conditions to test for such functions are computationally complex. It would be useful to explore into the possibility of obtaining simple necessary or sufficient conditions to test a finite sequence of autocorrelations for extendability so that its spectrum can be estimated. Also, non-square-domains of support for the autocorrelation (S_M treated is square and Δ is a higher dimensional boxed domain) need be considered.

REFERENCES

[1] B. W. Dickinson, 'Two-Dimensional Markov Spectrum Estimates Need Not Exist', *IEEE Trans. Information Theory*, 26, Jan. 1969, pp. 120–121.
[2] W. B. Arveson, 'Subalgebras of C*-Algebras 1', *Acta. Math.*, 123, 1969, pp. 141–124.

PROBLEM # 5

Rational Spectral Factorization

Consider the problem of spectral factorization of a multivariate rational spectral density function, $S(z_1, \ldots, z_n, z_1^{-1}, \ldots, z_n^{-1})$, having real coefficients, and which is strictly positive on T^n:

$$S(z_1, \ldots, z_n, z_1^{-1}, \ldots, z_n^{-1}) =$$
$$H(z_1, \ldots, z_n)H(z_1^{-1}, \ldots, z_n^{-1}),$$

where it is desired (but, this may not be possible, in general) that the spectral factor, $H(z_1, \ldots, z_n)$ be a real, rational, minimum phase function. We define a minimum phase function characterizing a 'causal' or positive cone filter to be one which is BIBO stable with a BIBO stable inverse; also, the support of the coefficients of the power series expansions about $(0, 0, \ldots, 0)$ of the function and its inverse are constrained to belong to the positive n–D cone (i.e. the first quadrant, when $n = 2$; the support can be analogously defined for weakly causal and asymmetric half-plane filters). Justify,† whether or not a minimum phase rational function must be devoid of nonessential singularities of the second kind on T^n i.e. whether or not the numerator and denominator polynomials of $H(z_1, \ldots, z_n)$ must be nonzero in $\bar{U}^n$. Since, it is well-known that a n–D $(n > 1)$ spectral density function almost never has a rational factorization, investigate the possibility of obtaining a set of necessary and sufficient conditions which a rational $S(z_1, \ldots, z_n, z_1^{-1}, \ldots, z_n^{-1})$ must satisfy so that a desired rational minimum phase spectral factor is obtained. Note that in [1.78], it has been shown that if the numerator and denominator of $S(z_1, z_2, z_1^{-1}, z_2^{-1})$ are strictly positive on T^2, then it is possible to express it as a quotient of two minimum-phase functions, each with support which is of infinite extent along one coordinate axis and of finite extent along the other coordinate axis.

† See *Conjecture 2* in Problem # 1.

PROBLEM # 6

Problem in Stabilization of Feedback Systems

(Contributed by J. P. Guiver)

It has been seen that if n, d are polynomials in $R[z, w]$, which are relatively prime and devoid of common zeros in $\bar{U}^2$, then there exist polynomials y, x, and ψ in $R[z, w]$ and an integer $N > 0$ such that

$$yd + xn = \psi^N$$

where ψ has no zeros in $\bar{U}^2$. The objective is to replace $R[z, w]$ by $Q[z, w]$, where Q is the field of rational numbers and investigate into the following problem.

Let $d(z, w)$, $n(z, w)$ be relatively prime polynomials in $Q[z, w]$, devoid of common zeros in $\bar{U}^2$. Find $q(z, w)$, $y(z, w)$, $x(z, w)$ in $Q[z, w]$, after proving their existence, such that

$$yd + xn = q$$

and

q has no zeros in $\bar{U}^2$.

For motivation and illustration consider the following simple example. Let

$$n(z, w) = z^2 + 2z - 1$$

$$d(z, w) = z - w + 2$$

which have common zeros for values (z, w) equalling $(-1 - \sqrt{2}, 1 - \sqrt{2})$ and $(-1 + \sqrt{2}, 1 + \sqrt{2})$. Clearly, via the procedure advanced in Chapter 3, together with the use of Gröbner basis,

$$\psi(z, w) = zw + (1 + \sqrt{2})(w - z) - (1 + \sqrt{2})^2$$

$$x(z, w) = 1$$

$$y(z, w) = -(z + 1 + \sqrt{2})$$

satisfy $yd + xn = \psi$ with $\psi(z, w)$ having no zeros in $\bar{U}^2$. However, ψ has non-rational coefficients.

By the continuous dependence of the zeros of a polynomial on its coefficients and because of the compactness of the unit bidisc, there exist intervals within which the coefficients of $x(z, w)$ and $y(z, w)$ can vary whilst not destroying the property that $xn + yd$ has no zeros in $\bar{U}^2$. In particular, we perturb the coefficients of $x(z, w)$ and $y(z, w)$ so that those are rational.

This argument will hold in general and proves the existence of $q(z, w) \epsilon Q[z, w]$. Let the perturbed polynomials be,

$$\hat{x}(z, w) = 1$$

$$\hat{y}(z, w) = -(z + \beta)$$

where the parameter β will be determined so that

$$\hat{x}(z, w)n(z, w) + \hat{y}(z, w)d(z, w)$$
$$= zw + \beta(w - z) - (1 + 2\beta)$$
$$\overset{\Delta}{=} \hat{q}(z, w)$$

has no zeros in $\bar{U}^2$. Invoking the tests for absence of zeros of $\hat{q}(z, w)$ in $\bar{U}^2$, i.e. (i) $\hat{q}(z, w) \neq 0$ in T^2, (ii) $\hat{q}(z, 1) \neq 0$ in $\bar{U}^1$ and (iii) $q(1, w) \neq 0$ in $\bar{U}^1$, it is easy to infer that a suitable choice for β is $\beta = 1$.

The problem posed here, with the polydomain of interest $\bar{U}^2$, could be extended easily to the case of an arbitrary compact polydomain.

PROBLEM # 7

Gröbner Bases, Polynomial Remainder Sequences and Decoding of Multivariate Codes

(B. Buchberger, E. V. Krishnamurthy, and F. Winkler)

Recently Krishnamurthy and Gregory [1], [2] have constructed codes (called Hensel codes) for a finite subset of rationals called the Farey rationals F_N satisfying

$$F_N = \{a/b = Q': gcd(a, b) = 1 \text{ and } 0 \le |a| \le N$$

and $\quad 0 \le |b| \le N\}$ where $N > 0$ is an integer.

If $N \le \sqrt{(p^r - 1)/2}$ then the mapping of the class of rationals F_N to the residue class of integers modulo m ($= p^r$ where p is a prime) then the mapping

$$| \cdot |_m: F_N \to I_m$$

where

$$I_m = \{|a/b|: a/b\epsilon F_N\}$$

can be made one-to-one and onto.

These codes $a \cdot b^{-1}$ mod p^r are called Hensel codes of Farey rationals and are equivalent to their finite segment p-adic expansions. The arithmetic with these codes turn out to be quite simple (similar to p-ary arithmetic) and the forward and inverse mapping from rationals to these codes turn out to be quite easy. Also, recently it has been shown [1] that the inverse mapping of the Hensel code to the corresponding Farey rational can be made using the extended Euclidean algorithm. This permits us to have a practical rational arithmetic system based on the Hensel codes.

Hensel's lemma and the Hensel codes turn out to be very useful for linear algebraic computations giving exact rational results [1], [2].

It is well known that there is a striking similarity between the algebraic structures of integers and the polynomials over a field. In fact these two structures are treated alike under the common algebraic structure – called the Euclidean or gcd domains. In view of this similarity, it is but natural to extend all the above concepts relating to the construction of Hensel codes of rational numbers to the rational polynomials over a finite field. Here, however, we need to deal with the more general class of Farey rationals called Padé rationals which are rational functions over a finite field (the coefficients are from a field) and the numerator and denominator degrees do not respectively exceed $(R - 1)$. Such a Padé rational is denoted by $P(R - 1/R - 1, F_p(x))$ a subset of $F(x)$, the rational polynomial functions over a finite field of characteristic p.

We then code

$$a(x)/b(x)\epsilon P(R - 1/R - 1, F_p(x))$$

as

$$a(x)b^{-1}(x) \text{ mod } x^{2R - 1}$$

which is the Hensel code [3].

The Hensel codes for these rational polynomials can then be used in a manner analogous to that for a rational number. This also provides us with a very effective tool for the symbolic manipulation and arithmetic of the rational polynomials over the integers by a suitable choice of p.

If $P(R - 1/R - 1, N, x)$ denotes the class of Padé rationals over integers where each coefficient is $\leq |N|$, then the choice

$$p > 2RN^2 + 1$$

enables us to construct Hensel codes for the practical problems.

The inverse mapping of these single variable rational polynomials can be obtained using the Euclidean algorithm. Thus a very practical rational polynomial arithmetic system can be devised and the use of Hensel's lemma again helps us to compute the solution for the linear algebraic problems and matrix inversion.

The above concept can further be generalized to multivariable rational polynomials [3] over a field and integers and the arithmetic can be performed in a similar way. This is useful for inversion of matrices whose entries are multivariate rational polynomials. However, the inverse mapping of the Hensel code to its equivalent rational polynomial cannot be realized by the Conventional Extended Euclidean algorithm since the multivariable polynomials are not Euclidean domains. The decoding has to be then based on the solution of a large matrix equation involving a Toeplitz matrix [4]. However, fortunately, it has been found recently that Gröbner bases [5] can be a very effective tool to decode the multivariable Hensel code. Naturally the algorithm for decoding the multivariable Hensel code will have several other applications – in multivariable Padé approximation [4], in the construction and decoding of multivariable Goppa Codes [6], and in the construction of matrix Padé approximants [4].

The examples below indicate the decoding of the Hensel code of a Farey rational, Padé rational and multivariable Padé rational. The proof of the algorithm for the Farey rational and single variable Padé rational is available in [1]. The proof of the multivariable decoding algorithm based on Gröbner bases has not been worked out in all the details so far.

Examples

1. Rationals

Let $p = 5$, $r = 4$; Hensel code of $\frac{10}{13} = 10 \cdot 13^{-1} \bmod 625 = 145 = .0401$.

The extended Euclidean algorithm:

i	q_i	625	0
		145	1

1	4	45	−4
2	3	10	13
3	4	5	−56
4	2	0	125

Decoding of Hensel code gives 10/13. (row 2)

2. Rational Polynomials over GF(p)

Let $p = 17$, $2R - 1 = 3$, $R - 1 = 1$;   Hensel   code   of
$(x + 1)/(2x + 1) = 1 + 16x + 2x^2$.

i	q_i	x^3 / $1 + 16x + 2x^2$	0 / 1
1	$9x + 13$	$4x + 4$	$8x + 4$
2	$9x + 12$	4	$13x^2 = 4x + 4$
3	$x + 1$	0	$4x^3$

Decoding of Hensel code gives

$$\frac{4x + 4}{8x + 4} = \frac{x + 1}{2x + 1}$$

3. Multivariable Hensel Code

Let $p = 7, 2 \cdot 2 \cdot (R - 1) = 4$; Hensel code of

$$\frac{5y + 3}{5x + 5y + 1} = 3x^3 + x^2y + 2y^3 +$$

$$+ 5x^2 + 6xy + y^2 + 6x + 4y + 3.$$

The algorithm [5] for constructing the Gröbner base:

i	p_i	t_i
1	x^4	0
2	x^3y	0
3	x^2y^2	0
4	xy^3	0
5	y^4	0
6	$3x^3 + x^2y + 2y^3 + 5x^2 + 6xy + y^2 + 6x + 4y + 3$	1
7	$4x^2y + 2xy^2 + 5y^3 + 2xy + 6y^2 + 3y + 4$	$2x + 6$
8	$2y + 4$	$2x + 2y + 6$
9	$5x + 4$	$4x^2y + 2xy^2 + 5y^3 + 6x^2 + 3xy + 4y^2 + 6x + 6y + 6$
10	5	$5xy^2 + y^3 + 4xy + 5y^2 + 6x + 4y + 4$

The column for the q_i's was omitted here, because there is no single quotient which can be associated with one step in the algorithm.

Decoding of Hensel code gives

$$\frac{2y + 4}{2x + 2y + 6} = \frac{5y + 3}{5x + 5y + 1}.$$

REFERENCES

[1] E. V. Krishnamurthy and R. T. Gregory, *Methods and Applications of Error-free Computation*, Springer-Verlag, New York, 1984.
[2] E. V. Krishnamurthy, 'Hensel's Methods in Linear Algebraic Computing – I', *Techn. Report CAMP* Nr. 83–27.0, Institut für Mathematik, Johannes Kepler Universität, Linz, Austria.
[3] E. V. Krishnamurthy, 'Hensel's Methods in Linear Algebraic Computing – II', *Techn. Report CAMP* Nr. 83–28.0, Institut für Mathematik, Johannes Kepler Universität, Linz, Austria.
[4] G. A. Baker and P. Graves-Morris, 'Padé Approximants', Parts I and II, Vols. 13, 14, *Encylopaedia of Mathematics and its Applications*, Addison-Wesley, Reading, Mass., 1981.
[5] B. Buchberger, 'Gröbner Bases: An Algorithmic Method in Polynomial Ideal Theory', Chapter 6, (this volume).
[6] R. J. McEliece, 'The Theory of Information and Coding', Vol. 3, *Encyclopaedia of Mathematics and its Application*, Addison-Wesley, Reading, Mass., 1977.

PROBLEM # 8

Invariance of Stability Property under Coefficient Perturbation

In system design, it is often necessary to preserve one or more characteristics of the system when system element values fluctuate about their respective nominal values. Investigations into the conditions for invariance of the multivariate polynomial positivity property under coefficient perturbation were undertaken in [1]. An interesting recent result of potential significance in the design of stable robust systems is due to Kharitonov [2]. Let

$$f(s) = \sum_{k=0}^{n} a_k s^{n-k} \qquad a_0 \neq 0$$

be a polynomial with real coefficients. (Kharitonov restricted $f(s)$ to be monic, i.e. $a_0 = 1$; however, his basic result is adaptable for non-monic polynomials and also generalizable to polynomials with complex coefficients.) Suppose that each coefficient a_k is allowed to assume any arbitrary value within the finite interval, $[\mathbf{a}_k, \bar{a}_k]$, i.e.,

$$\mathbf{a}_k \leq a_k \leq \bar{a}_k, \qquad k = 0, 1, \ldots, n.$$

Then, the following remarkable result holds.

Fact The polynomial $f(s) \neq 0$, Re $s \geq 0$ for arbitrary values of the coefficients in the finite intervals specified above if and only if each of the following *four* polynomials

$$f_1(s) = \mathbf{a}_0 s^n + \mathbf{a}_1 s^{n-1} + \bar{a}_2 s^{n-2} + \bar{a}_3 s^{n-3} + \mathbf{a}_4 s^{n-4} +$$
$$+ \mathbf{a}_5 s^{n-5} + \bar{a}_6 s^{n-6} + \ldots,$$

$$f_2(s) = \mathbf{a}_0 s^n + \bar{a}_1 s^{n-1} + \bar{a}_2 s^{n-2} + \mathbf{a}_3 s^{n-3} + \mathbf{a}_4 s^{n-4} +$$
$$+ \bar{a}_5 s^{n-5} + \bar{a}_6 s^{n-6} + \ldots,$$

$$f_3(s) = \bar{a}_0 s^n + \bar{a}_1 s^{n-1} + \mathbf{a}_2 s^{n-2} + \mathbf{a}_3 s^{n-3} + \bar{a}_4 s^{n-4} +$$
$$+ \bar{a}_5 s^{n-5} + a_6 s^{n-6} + \ldots,$$

$$f_4(s) = \bar{a}_0 s^n + \mathbf{a}_1 s^{n-1} + \mathbf{a}_2 s^{n-2} + \bar{a}_3 s^{n-3} + \bar{a}_4 s^{n-4} +$$
$$+ \mathbf{a}_5 s^{n-5} + \mathbf{a}_6 s^{n-6} + \ldots,$$

are nonzero in Re $s \geq 0$.

The proof of the above result depends on the following fact, which can be verified either by pure algebraic arguments as done by Garloff or by network realizability theory considerations, as done by Bose. See also [3], where, unfortunately, some of the proofs are incorrect.

Fact Let,

$$f(s) = f_e(s) + f_0(s)$$
$$g(s) = g_e(s) + g_0(s)$$

be polynomials with real coefficients where the subscript 'e' denotes 'even-part' and the subscript '0' denotes 'odd-part', i.e.,

$$f_e(s) = \frac{f(s) + f(-s)}{2}$$

$$f_0(s) = \frac{f(s) - f(-s)}{2}.$$

Then for any real $\lambda \epsilon[0, 1]$, the polynomial

$$f_\lambda(s) = (1 - \lambda) f(s) + \lambda g(s)$$

is devoid of zeros in Re $s \geq 0$, if
(a) $f(s) \neq 0$, Re $s \geq 0$, (b) $g(s) \neq 0$, Re $s \geq 0$, and (c) either $f_e(s) \equiv g_e(s)$ or $f_0(s) \equiv g_0(s)$.

Investigate into the possibility of obtaining the multivariate mathematical counterpart of Kharitonov's theorem for polynomial $f(p_1, p_2, \ldots, p_n)$ which is required to have no zeros in the closed right-half polydomain, Re $p_i \geq 0$, $i = 1, \ldots n$. Investigate also other feasible n-dimensional generalizations of Kharitonov's result, centering around those which find relevance in the context of studies of different types of stability for multidimensional systems.

REFERENCES

[1] N. K. Bose and J. P. Guiver, 'Multivariate Polynomial Positivity Invariance Under Coefficient Perturbation', *IEEE Trans. Acoustics*, Speech and Signal Proc., 28, Dec. 1980, pp. 660–665.

[2] V. L. Kharitonov, 'Asympototic Stability of an Equilibrium Position of a Family of Systems of Linear Differential Equations', *Differential Equations*, Plenum Publishing Corporation, 14, 1979, pp. 1483–1485.

[3] S. Bialas, 'A Necessary and Sufficient Condition for the Stability of Interval Matrices', *Int. J. Control*, 37, 1983, pp. 717–722.

PROBLEM # 9

Stability of discrete systems under coefficient perturbantion

A polynomial which has all its zeros within the unit circle is a strictly Schur polynomial. Such a polynomial occurs in the stability studies of linear time-invariant digital filters. Investigate the scope for obtaining the counterpart of Kharitonov's result stated in Problem # 8, for the case of a set of strictly Schur polynomials. Are four polynomials formed in a certain way, from the values of the upper and lower bounds specified for each coefficient, necessary and sufficient to test for the strictly Schur property of a set of polynomials whose coefficients are bounded from above and below by the specified real numbers? If that is not the case, justify your conclusion and explore the feasibility of obtaining an optimum test for the strictly Schur property of a set of polynomials, whose coefficients may take arbitrary values within specified intervals.

PROBLEM # 10

Stability of Matrix Polynomials Under Coefficient Perturbation

Let $H(s)$ be a matrix whose entries are rational functions in s with real coefficients. If $H(s)$ is represented by a typical matrix fraction description, $H(s) = A^{-1}(s) B(s)$ (where $A(s)$ and $B(s)$ are relatively left prime polynomial matrices with det $A(s) \neq 0$), then $H(s)$ characterizes a stable system if and only if det $A(s)$ has all its zeros in the open left half-plane, Re $s < 0$. Of course, $H(s)$ could also be represented by a typical right matrix fraction description, $D(s) \; C^{-1}(s)$ (where $C(s)$ and $D(s)$ are relatively right prime polynomial matrices with det $C(s) \neq 0$) and then for stability it would be necessary and sufficient that det $C(s) \neq 0$ in Re $s \geq 0$. For the sake of brevity, a $C(s)$ satisfying det $C(s) \neq 0$ in Re $s \geq 0$ will be called strictly Hurwitz. In attempting to provide a generalization of the result linking strictly Hurwitz polynomials (i.e. polynomials devoid of zeros in Re $s \geq 0$) and reactance functions, Anderson and Bitmead obtained certain interesting results [1] for determining whether or not all the zeros of the determinant of the polynomial matrix belong to the open left-half plane in terms of whether or not the reactance matrix property is satisfied by a derived rational matrix. It should be noted, however, that the derivation of this rational matrix from the specified polynomial matrix is not straightforward like in the scalar case. Possibly exploiting the results in [1], obtain conditions under which the linear convex combination of two strictly Hurwitz polynomial matrices will be strictly Hurwitz. Subsequently, investigate the scope for obtaining the polynomial matrix counterpart of Kharitonov's result, stated in Problem # 8, when the coefficients of the elements of the polynomial are bounded from above and below by specified real numbers.

REFERENCE

[1] B. D. O. Anderson and R. R. Bitmead, 'Stability of Matrix Polynomials', *Int. J. Control*, 26, 1977, pp. 235–247.

Index